Ideas & Actions

A HISTORY OF THE
HIGHWAY RESEARCH BOARD 1920-1970

National Research Council
National Academy of Sciences

Ideas & Actions:
A HISTORY OF THE HIGHWAY RESEARCH BOARD,
1920-1970

ISBN 0-309-01830-7

Price $12.00

Available from
Highway Research Board
National Academy of Sciences
2101 Constitution Avenue N.W.
Washington, D.C. 20418

Contents

Foreword

The Executive Committee of the Highway Research Board authorized a commemorative history of the Board to be written in connection with the observance of its 50th anniversary. This version provides the essential facts in full. A condensed version, portraying only the more significant highlights, will be distributed at the anniversary meeting in January 1971.

By and large this narrative was written by two authors. Begun by the late Fred Burggraf, who served as Director of the Highway Research Board from 1951 to 1964, it was completed by M. Earl Campbell, who served the Board in various capacities from 1947 to 1967. Chapters 7 and 9 were written by other staff members.

An interpretative history was attempted. It sketches the reasons for the Board's creation, its role, and its impact on highway transportation, administration, and technology over the years. It describes the Board's relationship with the National Academy of Sciences and the National Research Council and pays tribute to the encouragement and wise counsel that have always been available from those distinguished parent organizations. The narrative describes the evolution of ideas and actions of the highway research community. The pooling and sharing of these experiences through 50 years have had a dramatic effect on today's world.

This story is offered as a testimonial to the founding fathers and to the host of men and women who have participated in the work of the Board during its first 50 years. More than just commemorating the past, this chronicle provides guidance and challenge for the future.

Many of the records and publications of the Board used in this history—a rich legacy from the past—are shelved in the library of the Board and are available to anyone who may wish to pursue them in more detail.

William N. Carey, Jr.
Executive Director

Preface

This is a story of the Highway Research Board and its role during the period from 1920 to 1970. The occasion of this history is the 50th anniversary of the Highway Research Board. It is a tribute to the men who have brought the Board to its honored position. It serves as a testimony to the Board's half-century of service in the advancement of transportation technology and use. Its chief value, it is hoped, will be more than a grace note to honor the past. It is intended to unveil the legacy of knowledge developed and left to us by pioneers in the concept of pooled sharing of new discoveries in transportation technology. In addition, this history is intended to reveal the *raison d'etre* of the Board.

It has been a refreshing experience to become immersed in the dialogues of this half-century and to find such a fascinating flow of ideas. The questions and the problems of 50 years ago were quite like those of today. Even in 1919, upon hearing the problems proposed, a researcher said, "I see nothing new in these. We were working on them 25 years ago." Final solutions are seldom found. Understanding what people thought 25, 50, or 75 years ago, the actions they took, and the consequences of their actions provides a clue to what can be done successfully today and what tends to fail.

It would not be too difficult to enumerate the accomplishments of the Board, but to treat its history as a series of discrete chronological incidents is to miss the more significant point—its role and the reason for its unique existence. This account then accents concepts, questions, problems, solutions, and the consequences of ideas put to work. It is necessarily an abbreviated story of activities and accomplishments, but the ideas that fashion activities of the future cannot be cut short—they will continue.

This narrative was written primarily by two authors. Fred Burggraf, whose life was vitally linked with the Board from 1929 to 1966, began the writing. His death on January 25, 1966, ended his authorship. His signature is seen in Chapters 2, 5, and 6. But he put more than his signature to the chronicles of the Board through the many years he served—13 as Director. Many of the enduring ideas and many of the special projects of the Board were his. He not only wrote history, he made it.

HRB Director Carey, with Staff Economist Kenneth Cook, wrote Chapter 9. Mr. Carey, because of his involvement with the Board's special projects, edited and wrote most of Chapter 7.

Otherwise the job of completing the task begun by Mr. Burggraf has fallen to me. It has been a fascinating as well as difficult assignment. I began work with the Board in 1947 and have had the pleasure to be associated with and work under the leadership of Directors Crum, Burggraf, Mickle, and Carey. I knew quite well Mr. Upham, who preceded Mr. Crum as Director. This association has inspired my writing. But my association with these men and a host of other people who participated in the work of the Board, together with my own deep involvement in the activities of the Board through a score of years, has made it difficult to write objectively.

Any author of the history of the Highway Research Board would find a tremendously difficult job in selection of materials. It is a formidable assignment to attempt to select and evaluate the important out of a 30-foot shelf of HRB publications, out of its 7,500 technical reports by 5,000 authors, out of 30,000,000 words.

It would take a full year just to scan the publications. And many words significant to the Board's history appear in publications of other organizations. It is clear then that an amalgam of these words is not the desired end. It is also clear that this narrative cannot summarize findings from a national highway research effort that is now approaching $50 million a year, and that has cost $500 million since 1920.

It is not too hard to portray an institution's physical growth by its increasing budget, its expanding staff, and its accelerating activities—and these are worth portraying—but a more difficult portrayal and one that the historian must attempt is that of the growth of service and the diffusion of worthy ideas in useful form. These are the resources that make a history splendid.

My sources of information have been many. At hand was an assembly of much information dug out of the records by Mr. Burggraf. I have searched the literature from many organizations. I have also been fortunate in being able to interview many of the people whose participation in the Board's affairs has spanned all or most of the Board's half-century of service. In my 20 years with the Board I enjoyed stimulating dialogue with many of the people who have made the Board. The close association with five of the seven directors and with all of the professional staff of the Board has produced osmotic effects helpful to this narrative.

During these 20 years visits and exchanges of letters with key people in all state highway departments and with professors in highway-related activities in 120 universities, along with visits to city planners and traffic engineers in many cities, have also had a profound effect. Added to this has been the exposure to ideas and actions in department and committee meetings and association with highway specialists here and abroad. This has been my inheritance. In this story of ideas and actions I am trying to share with the reader some of these valued gifts.

In presenting this story in self-contained chapters there has been some necessary reiteration. This seemed the best way to handle a story of such complexity.

Writing this history has been a prized privilege. My appreciation is extended to the Executive Director of the Board and to the host of people who helped.

One must read between the lines of poetry and history to behold the image that cannot be contained in words.

M. EARL CAMPBELL
January 1, 1971

Acknowledgments

The Highway Research Board gratefully acknowledges the assistance given the authors of this history by many individuals and organizations. In particular, the Board acknowledges with sincere appreciation the contribution of the West Virginia Department of Highways in affording time to Mr. Campbell to complete the work he assumed upon the death of Mr. Burggraf.

The history was reviewed and many thoughtful suggestions were made by the members of a special committee chaired by J. Gibbons and including J. E. Buchanan, C. D. Curtiss, E. H. Holmes, F. W. Hurd, L. Jordan, G. D. Kennedy, K. B. Woods, and D. G. Mickle.

Finally, the full manuscript was edited by HRB staff members Marjorie Moore, D. H. Buswell, and Stephen Montgomery.

1

Prologue: The Milieu

The formation of the Highway Research Board followed a necessary and eventful prologue. It grew from a stream of converging problems. These problems issued from a spectacular development some 30 years earlier—the invention of a practicable motor vehicle. From this unique development there emerged accelerating forces and pressures that brought about the concept of a highway research board. The story of this prior period is the prologue—the story of the generation of preparation.

The Gay Nineties

In 1890 census enumerators counted 62,947,714 people and nearly 20 million horses and mules in the United States. An estimated 7.5 million carriages of countless varieties carried passengers, and 5 million wagons drew freight. One and one-half million new horse-drawn vehicles were produced during the year, about 60 percent of these for "movement of people." A million bicycles were sold that year. The gross national product (in 1929 dollars) was nearly $27 billion. (Seventy-five years later highway transportation contributed $90 billion to the GNP.) There were only 200 miles of high-type paved rural highways.

There was a freshness and freedom that gave an exhilaration to the Nineties—perhaps more in one's memory than in real life. Yet there was a pervasiveness to the glamour of the new "bicycle age." By the mid-Nineties 4 million Americans were happily exploring their neighboring vales and hills on bicycles.

Before this decade began, bicyclists had joined to organize the League of American Wheelmen and to inaugurate the Good Roads Movement in America. The efforts of the League began to bear fruit in the enactment of local road aid laws—seeds of the road renaissance. Between 1878 and 1900 seven state highway departments were created, the District of Columbia Highway Department having been established in 1878.

Although a German-made Benz automobile had been exhibited at the World's Fair in Paris in 1889, the automobile was hardly more than a gleam in the eye of Americans until 1892, when the first American automobile was built by Duryea. Except for its stirring of vision, the impact of the motor vehicle during this decade was negligible. Fewer than 8,000 automobiles were made between 1892 and 1900. During this period nearly 8 million surreys and a comparable number of bicycles were made. But the vision had been implanted, and it dimmed the glory of the surrey and the bicycle.

There were other significant events toward the end of the 19th century. In 1893, as a consequence of the Good Roads Movement and in acknowledgment of the technological problems involved in road development, the Office of Road Inquiry (the forerunner of the Bureau of Public Roads) was organized. This organization immediately began a program of inquiring into the best methods of road building. It engaged in testing, experimenting, and investigating. It began to teach through dissemination of information and "seeding" with demonstration roads. Yet it was not until 1895 that four automobiles were registered.

Another spectacular event that gave tremendous impetus to the road program was the inauguration of the Rural Free Delivery of United States mail in 1896. "Those interested in development of good roads . . . recognize that good roads and rural mail delivery are necessarily closely connected," said the annual report of the Post Office Department in 1899.

By the end of the decade, because of the impact of all these events, experimental sections of portland cement concrete and macadam surfacing had been tried. Brick had been used in street construction. Sand-clay roads were gaining in popularity in some rural areas because they were economical and effective for the traffic of that day. Dust palliatives were coming into use.

September 13, 1899, marked the first fatality from an automobile accident—an ominous portent of the destructive aspect of the motor vehicle. Nevertheless, the most important event in the record of transportation history was the genesis of the automobile. As the century ended this powerful factor was a reality, and the world—almost imperceptibly then—had begun to shrink.

The Twentieth Century Begins: 1900-1909

The United States began the 20th century with a population of 75,994,575, comprising nearly 16 million households. Almost 6 million farms were the home of 60 percent of the population. There were about 20 million horses and 5 million mules. An estimated 9 million carriages and 6 million wagons traveled an estimated 12 to 15 billion vehicle-miles that year (an average 2¼ to 2¾ miles per day per vehicle). Trips exceeding 12 to 15 miles in length were infrequent. Horse-drawn passenger vehicles traveled at 6 to 10 mph and wagons at 2 to 3 mph. Although bicycle production declined after 1897, 2 million were made that year. An inventory of bicycles in 1900 might well have shown 10 million.

In 1900, 8,000 automobiles were registered, and 4,192 were built during the year. There were still barely 200 miles of rural high-type pavement, although 135 cities reported 15,000 miles of paved city streets. A survey made in 1904 recorded 154,000 miles of "surfaced" roads and streets, but a surfaced road was not necessarily a paved road and usually was not.

The world in 1900 was relatively peaceful. The United States had emerged from the Spanish-American War in 1898 with its position enhanced, and by 1900 was becoming an industrial power among the nations, exporting more than it was importing in foreign commerce. The economy was flourishing. The GNP reached nearly $37 billion, growing a little faster than the population. The country faced the new century with a sense of challenge and a sense of strength.

The horse and buggy age continued and reached its peak during the first decade of the century. Automobile production began to match production in the carriage industry by 1905, with the substitution approximately on a one-to-one ratio for about 5 years. Carriage production in 1909 fell nearly 10 percent, the drop being approximately equal to the number of automobiles produced that year.

This changing environment produced other events. In 1902 the American Road Builders' Association was formed and began work to aid the road-building industry and to augment the effort for road development. The

American Automobile Association was formed in the same year and began an organized effort for improved highways and personalized service to the motorist.

There was little directional signing on rural roads. There were few places where a motorist could buy gasoline and oil, and few places, except city hotels, where a motorist could find a place to eat or lodge. Despite this situation a motorist in 1903 drove from the Pacific Coast to the Atlantic Coast in 63 days (44 days running time). Many people had made cross-country trips before the end of the decade.

The number of automobiles increased to 304,950 by 1909. Deaths resulting from auto accidents also increased. The death toll rose to 40 per year for each 10,000 registered vehicles. (This had dropped by 1968 to 5.41 fatalities for each 10,000 registered vehicles.) There were 16 deaths per 100,000 population (in 1968 this figure was 27.6 deaths per 100,000 population), and there were more than 20 fatalities per 100 million vehicle-miles (in 1968, 5.47). Also in 1909, the peak year for travel by horse power, there were some 21 million horses and mules at work traveling 13 billion miles annually, and there were 30 fatalities for 100 million horse-drawn miles.

The need for instruction in highway engineering, if road development were to match demand, was recognized as soon as motor vehicle production began to mount. By 1909 half of the engineering colleges had instituted highway courses, and several textbooks had been written on highway construction and testing of materials. But the technology used was that developed in building roads for horse-drawn vehicles. Not for another 10 years was a road design compatible with the motor vehicle introduced, and then the geometric standards were adapted from railroad practice. The Office of Road Inquiry, which in 1905 had consolidated with the Division of Tests of the Bureau of Chemistry to form the Office of Public Roads, had early noted the need for more extensive highway research, but the urgency was not generally recognized in this first decade.

If it can be assumed that the motor vehicle was the most important transportation development of the Nineties, it follows that the eager acceptance of the motor vehicle was the significant characteristic of the period from 1900 to 1910. The horseless carriage had an appeal far greater than had been imagined possible. People had quit crying "Get a horse," and had begun buying automobiles as fast as they were produced, often paying five times as much as for a carriage. (Even then an automobile meant freedom and opportunity and status.) Consequently, the vehicle tally began to move swiftly from the one-vehicle-to-five-persons ratio of 1900 toward the one-to-two ratio now near at hand.

Early Pressures: 1910-1914

By 1910 the population had increased to 91,972,266, and the GNP had reached $55 billion, approximately $600 per capita. The number of horses and mules still exceeded 20 million, but people began to realize that the horse was going out of fashion. The number of horse-drawn vehicles was on a rapid rate of decline but still exceeded 15 million. Motor vehicle registration increased; 458,377 autos and 10,123 trucks were added in the year's production. The bicycle population was still exceedingly large.

More than 200,000 miles of road were "surfaced," yet stretching over our vast continent there were more than 2 million more miles of dirt roads awaiting surfacing. These were about 12 to 14 ft wide with grades often in excess of 10 percent and with turns and curves built for negotiation at a maximum speed of 5 mph. Cost of freight transportation averaged in excess of 10 cents per ton-mile, and the average load carried by wagon was about 1 ton.

In 1912 Congress authorized $500,000 for an experimental program of rural post-road construction to determine savings in rural mail operations as a result of road improvement.

In 1914 (November 11 in Atlanta and December 12 in Washington, D.C.) the American Association of State Highway Officials was organized with 18 states represented. (By 1920, 31 states and the Bureau of Public Roads participated as members, and in 1922 all states were represented.) This organization provided a great new thrust for highway improvement and for encouragement of experimental work. Although there were no officially designated units of highway research in the state highway departments at this time or through 1920, it is known that a number of states were making modest efforts in road experiments.

In 1914 the growing production of autos almost equaled the rapidly declining production of horse-drawn carriages (538,000 autos to 548,000 carriages). Altogether, in 1914 there were 1,664,003 automobiles and 99,015 trucks registered. Mixed traffic was increasing and becoming a growing problem to safety and highway capacity—annual vehicle-miles by motor vehicle were about equal to annual vehicle-miles by horse-drawn carriage at this time.

Although 50,000 more miles of roads and streets had been surfaced in the preceding 5 years (largely by local authorities), the rate of improvement did not satisfy owners of motor vehicles. There was a growing clamor for instant improvement of streets and roads.

The problems that began with the use of the motor vehicle were multiplying in geometric progression as the years passed. To find solutions to these problems, highway research was becoming imperative.

The Conjoining Forces: 1915-1919

Some notable events during the 5 years of swift change from 1915 to 1919 added tremendous impetus to the burgeoning problems in highway transportation and to the evolving concepts that led to the creation of the Highway Research Board.

First, there was the impact of the Federal-Aid Road Act of 1916 that provided federal-aid funding for construction of highways. The funding was conditioned by two provisions: that the aid be matched by the states and be administered by state highway departments. This resulted in the establishment of highway departments in all 48 states by 1919, an increase from 30 states in 1916. The Act provided apportionment of $75 million to the states (a significant amount in 1916). A most profound result of this funding was the partnership endeavor effected between the federal and state governments in highway development. This partnership was to exert a mutual influence in the coming years for improvement in planning, design, and research.

Second, there was the impact of World War I, which involved the United States from April 6, 1917, to November 11, 1918. Among the results of this involvement was a greatly stepped-up production and use of motor trucks, particularly the hauling of loads of greater magnitude and frequency in the war effort than had ever been witnessed before in the United States. Roads were expendable, and hard use resulted in damage to many surfaces built for the prevailing 1- to 2-ton loads carried by wagons. Loads of 5 tons were becoming common and highway engineers recognized that sights must be raised because such loads of 5 tons (up to 12 tons gross) might be anticipated in the near future as the customary average load. Thus, in a sense, the war provided in size and weight concepts a framework for economic progress, breaking through barriers of inertia that might have persisted many years without its impact.

One of the big jobs after the war was road construction and new construction from accelerated road funding. This brought highway administrators, truck manufacturers, and legislators face to face with the problem of load restrictions. To focus on the problem of truck loading, a joint meeting of AASHO and the Highway Industries Association was arranged in Chicago on December 9, 1918. Feasible legislative policies on size and weight and matters of regulation and enforcement were discussed. Out of this conference, attended by highway officials and motor truck manufacturers, came the recommendation that was to shape the structural and geometrical design of highways for many years—the recommendation for a maximum 14-ton gross load and a maximum 800-lb load per inch of solid tire width.

It becomes quite clear from the literature of the day that there was no economic or experimental basis for the establishment of load restrictions. The 800 lb per inch of tire width limitation was based on a recommendation of the tire manufacturers as the maximum economic tire loading. At that time 14 in. was the maximum width of tire produced. Thus, at 2 wheels per axle, $2 \times 14 \times 800 =$ the 22,400-lb maximum axle load that found its way into the load restriction in eight states. These states allowed gross vehicle loads of about 28,000 lb.

Another group of states permitted a maximum axle weight on the order of 16,000 lb distributed at 800 lb per inch of solid tire width (with some states allowing less than 800 lb) with no maximum gross load specified. A third group specified only maximum gross load and maximum load per inch of tire width.

Although knowledge of stress distribution in layered systems was recognized as a basic need in structural design of highways, no research along this line had yet been done. Now that state legislatures were enacting weight restrictions and providing construction funds, there was a convenient starting point for road tests. In Illinois, for example, where the law now provided a maximum axle load of 16,000 lb distributed as 800 lb per inch of tire width and where a $60 million bond issue for road construction had been authorized in 1918, the state highway department prepared to undertake the Bates Road Test to determine the required structural design of pavements to support the allowable loads (see Chapter 7).

Another result of World War I that helped shape the environment for research was a change in the attitude toward research. People had seen the contribution made by research to the war effort, and it was becoming accepted as a worthy and necessary endeavor in the pursuits of peace. President Wilson helped mold this thinking by requesting that the National Academy of Sciences continue the National Research Council after the war and undertake new assignments.

There was still another result of the war that influenced highway development and its related research endeavor. Men returning from the war had had a taste of travel. In enforced associations with other people, these veterans had experienced the advantages of social exchange. They had seen the world and made new friends and were now tired of the isolation that a mud road enforces. Their minds had been emancipated from parochial ruts, and now with the automobile available they wanted to be physically free as well.

The impact of the exploding registration of motor vehicles on social life and the economy* and the enormous demand for roads adapted to the motor vehicle and usable at all times had significant effects in this period on highway law, financing, and development. The craze for driving was the stimulus to action, the catalyst to problem-making. These problems became the stuff of research.

In the years from 1915 through 1919, the funds spent on state-administered highways exceeded a half-billion dollars.

In 1919, with the war over, many miles of road had to be rebuilt and many new miles had to be built. In every state the urgent question was raised, "How?" Other questions that needed answers included: For what vehicle load? For what vehicle size? For what vehicle speed? For what traffic volume? For what traffic composition? By what means of financing? How were funds to be apportioned? How could directions be given to travelers? How could pavements be conserved?

The list of questions was endless. Such questions stemmed from growing problems, such as:

Mixed traffic
Increasing frequency and magnitude of loads
Increasing speeds
Growing accidents and accident costs
Growing vehicular congestion
Bridges inadequate in number and structural capacity
Small mileage of surfaces adapted to motor vehicles
Large mileage of inferior roads
Small budgets for the demands
Small number of highway engineers

* A pre-World War I estimate indicated that it cost more than $100 per month to maintain and operate one horse and one wagon, itemized as follows (based on a 17-mile trip per day):

Horse keep and care	$ 35.00
Horse cost and depreciation	5.00
Wagon cost and depreciation	4.50
Driver's wages	60.00
Wagon and equipment repair	7.50
Total cost per month	$112.00

Horse cost $240 (life 5 years); wagon cost $250 (life 7 years); harness cost $40 (life 2 years). Cost of hauling was thus more than 25 cents per mile.

A writer estimated that a horse eats 12,000 lb annually and that "the acreage devoted to keeping 24 million horses at 5 acres per horse is more than that sufficient to feed the entire human population of the country."

General Motors Corporation advertisements in those years stated "one GMC and one driver displaces 16 horses, 4 drivers, and 4 wagons." Trucks could travel 12 miles per hour, 50 miles a day and could operate for at least 10 cents per mile less than a horse and wagon.

At this time men of vision had already anticipated research needs as they recognized the potential conversion to motor vehicles. Highway research had gotten under way. Scattered across the continent, modest studies—perhaps not always dignified by the name of research—were being started informally in highway departments, colleges, and highway-related industries.* Prominent in highway research activities during these years was the Bureau of Public Roads, which had been seriously engaged since about 1915 in highway research at Arlington Farms (the agricultural experimental farm then operating opposite Roosevelt Island along the Potomac River in Arlington, Virginia). These research efforts, while growing, were scattered, isolated, decentralized, and certainly uncoordinated.

Before the year was over (October 31, 1919) the Division of Engineering of the National Research Council authorized the creation of a Highway Research Committee "to coordinate and assist highway research work now being conducted by the United States Bureau of Public Roads, by the state highway departments, by manufacturers, research departments, and commercial laboratories."

But the time for the Highway Research Board was not yet at hand. Another full year was required—a year of deliberation and review.

The Beginning of a New Decade: 1920

The United States entered the 1920's with a population of 105,710,620 in some 24 million households. There were 6,448,343 farms, but only 30 percent of the people lived on farms. The GNP (in 1929 dollars) was $73 billion, increasing much faster than the population in the previous decade. Although there were still more than 25 million horses and mules in the United States, the estimated number of horse-drawn vehicles in this decade had declined from more than 15 million to about 9 million. During 1920 the motor vehicle, with its accelerating registration, matched the estimated number of horse-drawn vehicles: the automobile registration for 1920 increased to 8,131,522 and motor truck registration increased to 1,107,639.

* A review of state highway department and university records showed total expenditures for highway research in 1920 to be about $175,000 by state highway departments (Alabama, California, Connecticut, Illinois, Iowa, Kentucky, Massachusetts, Michigan, New Hampshire, New Jersey, New York, North Carolina, North Dakota, Oregon, Ohio, Pennsylvania, Texas, Utah, Virginia, Vermont, West Virginia, and Wisconsin) and about $150,000 by colleges and universities (in Alabama, Colorado, Georgia, Idaho, Illinois, Indiana, Iowa, Kansas, Maryland, Massachusetts, Missouri, Michigan, Nebraska, New Hampshire, New Jersey, Ohio, Pennsylvania, South Carolina, South Dakota, Texas, and Washington). Economic research at that time, including traffic, finance, and road costs, was supported at about $15,000 of the total $325,000.

Together, horse-drawn and motor vehicles on the roads and streets totaled more than 18 million, not counting motorcycles, which numbered 241,000, plus bicycles numbering in the millions.

Thus during 1920 there was in the United States (as there is now 50 years later in developing countries) an extreme mix of incompatible components in the traffic stream—autos, motor trucks, carriages, wagons, motorcycles, bicycles, saddle horses, and pedestrians—with droves of sheep and herds of cattle often preempting the road. Congestion and accidents were high. It has been estimated that the fatality rate per 100 million miles for horse-drawn vehicles was about 30 (for an estimated 10 billion miles of travel in 1920). The fatality rate for the motor vehicle was about 20 per 100 million vehicle-miles (for an estimated 60 billion miles of travel that year).

Conflicts, both physical and psychological, were inevitable in this period of swift and difficult transition from horse to motor. Travel speed of horse-drawn carriages ranged from 6 to 10 mph and for loaded wagons from 2 to 3 mph, while speeds for autos ranged from 20 to 35 mph and for loaded trucks from 10 to 20 mph. The differential speeds occasioned frequent and difficult passing maneuvers, and consequent accidents. Blind curves, narrow roads, inadequate vehicle lighting, few stop signs or signals, scarcely any fixed lighting in rural areas and untrained drivers induced physical conflict. Highway structures (bridges and surfaces) built for maximum gross loads of 3 to 5 tons were stressed beyond design capacity and failed accordingly.

In short, the problem was to determine how to match design to performance requirements. In other words, how could the road be made economic and adequate for community, user, vehicle, and abutter in all of the various demands of each?

The Unique Idea

The increasing research effort was scattered, isolated, desultory, and decentralized. Decentralization was generally favored and endorsed. Decentralization could enlist the interest and aid of more people, it could encourage more individual initiative, it could provide for the important investigations of peculiar local problems, it could institute cooperative research activities in its own locale, it could spread the efforts necessary on nationwide problems, it could profit by blending the viewpoints from many regions and institutions, and finally it could and would use the findings from its own efforts in its own operations. There were a few who favored centralization of research, but the proponents of decentralization had their way.

At the same time it was recognized that decentralization had left a communication problem in a day when just the mechanics of communication were difficult. If decentralization was to be the way of highway research in this vast country, there must be a clearinghouse—a center of communications for the growing number of autonomous, isolated research units. Determining how to foster decentralization with its observed potentialities, yet accomplish broader objectives in highway research, was given considerable thought. The targets were national as well as local, and avoiding duplication and waste were of prime importance. It was apparent that there must be continuing motivation and mechanics for correlation, and that there must be an articulation of the need and the benefits of a clearinghouse, and a way found to make it work through practicable means of pooling and sharing information, exchanging ideas, interpreting discoveries, and stimulating the total effort.

As one scans the literature of that day and listens to the remaining few of that original group of pioneers, it becomes evident that they perceived, perhaps more clearly than we, that there is a mutual dependency and mutual support in the activities of research, education, and application. They saw these as one process rather than three separate processes. They recognized the input and output among each of these interdependent efforts. The highway official, the educator, and the researcher had a conscious vital stake and a common urgent interest in the total effort, more so than we do now a half-century later when the excitement is dimmed and each of these efforts has become compartmentalized within great and diverse institutions.

Origin of the Concept

One cannot single out one individual and say that he alone conceived the master idea for the Board. Indeed, there is evidence, as one sifts through the early papers and pronouncements, that the idea evolved through accretion. The committee created in 1919 was pre-HRB in concept. Nevertheless, certain men in strategic positions who played a principal part in forging the 1920 pattern can be identified.

Anson Marston, at that time Dean and Director of the Engineering Department of Iowa State College, was also a member of the Iowa State Highway Commission. Dean Marston enunciated his deep interest in a coordinated national highway research program. At the Fifth Annual Meeting of the American Association of State Highway Officials in Louisville in December 1919, he presented a paper entitled "A National Program for Highway Research." He was also a member of the organizing committee for the Board and became the first Chairman of the Executive Committee.

Thomas H. MacDonald, the catalyst and prime mover, had been a student of Dean Marston's and had worked as his graduate assistant. State Highway Engineer in Iowa until 1919, when he was appointed Chief of the U. S. Bureau of Public Roads, Mr. MacDonald was a strong proponent of decentralization and of a partnership approach to highway development by the Bureau and the states. While on the organizing committee he promoted and supported the aims and activities of the Board. Through BPR funding he sustained the Board until it obtained firm footing many years later when the state highway departments began their systematic support.

There were others: A. N. Talbot, Professor of Engineering, University of Illinois, and past President, American Society of Civil Engineers; Clifford Older, Chief Engineer, Illinois Highway Department, and President, Mississippi Valley Conference of State Highway Departments; and Thomas R. Agg, Chairman of the AASHO Committee on Investigations and Tests and Testing Engineer of the Iowa State Highway Commission. These men and others who helped shape the Board in its creation and mold it in its early years are named in the next chapter.

One other person, Comfort A. Adams, deserves special mention here because of his sympathetic interest and strategic position as Chairman of the National Research Council's Division of Engineering. Dr. Adams, Lawrence Professor of Engineering at Harvard University, was then devoting full time to the work of the Division. Dr. Adams emphasized the importance of cooperation and coordination in research work, advising that "the purpose of the National Research Council is to bring scientific methods to bear on the solution of practical problems of today."

These men probably developed the general concept of a highway research board in prior dialogues, and then, in concert with a larger group on November 11, 1920, put thought into action. What they had in mind was a simple idea of service—service to be achieved and maximized by linking mutually supporting, mutually dependent activities of research with a minimum of cost brought about by a correlation of ideas and activities. Their objectives were:

To identify feasible problems for highway research to serve the highway industry's needs

To formulate a statement of nationwide highway research needs to serve the public interest

To stimulate efforts in highway research, education, and application

To correlate highway research activities to make them more conclusive, efficient, and effective for highway departments, universities, and highway industries

To present and discuss problems and ideas through group assembly

To pool and share research resources, data, and new knowledge

To collect and disseminate highway research findings through an information service

To conduct sponsored research when feasible

To administer and mediate—not manage

The mechanics for putting the idea into practice included committees, publications, annual forums, and personal contacts.

The objectives enunciated in 1920 are still prime objectives today: to stimulate, correlate, discuss, and disseminate. In the growth of the Board from one staff member in 1920 to over 100 in 1970 without a change in basic goals, there is dramatic evidence of the simplicity, soundness, and desirability of the original objectives. The vision encompassed research on the vehicle, the road, the road user, and their environment together with their interrelationships. The mission has not changed; only the mechanics have changed to keep up with the advancing technology in research and communications. The service concept has not changed; only the acceleration and coverage have changed. The service was to help provide facts pertinent to decision-making but not to recommend decisions.

The time to invest concepts with a corporate body had arrived and was planned for November 11, 1920. But how could such an idea become a corporate being and survive? It would need a climate providing the freedom to pursue truth without interference of political pressure or proprietary interest, to publish the truth so far as truth is known, to serve without favor, and to be responsible to its mission.

Should the organization be formed as an independent body, be attached to a federal bureau, or be a unit of some other highway-related institution? Fortunately, the concepts of the organizers matched the policies of the National Research Council of the National Academy of Sciences, and there the organization found a home.

1 Horsepower ratings for hauling equipment in this Ohio setting meant exactly that, although the mechanical shovel was already becoming commonplace when the photograph was made in 1914. The equipment is rough-grading a highway near Linville.

2 Sixty years ago the science of producing portland cement concrete was in its infancy. This early rotary mixer was photographed in 1909 on the grounds of Cornell University in Ithaca, New York.

1

2

2

Creation of Organization and Functions

The Parent Organization

The Highway Research Board was organized in 1920 as an agency of the Division of Engineering, one of the functional divisions of the National Research Council of the National Academy of Sciences. A brief statement on the origin and functions of each of these organizations follows.

The National Academy of Sciences was established in 1863 under a congressional charter signed by President Lincoln (see Appendix A). The Academy is a private, nonprofit organization of scientists, dedicated to the furtherance of science and its use for the general welfare. Empowered to provide for all activities appropriate to academies of science, it is also required by its charter to act as an adviser to the federal government in scientific matters. This provision accounts for the close ties that have always existed between the Academy and the government, although the Academy is not a governmental agency.

The National Research Council dates back to April 1916 when the National Academy of Sciences offered to President Wilson the services of the Academy in the interests of national security and welfare and preparedness for the emergency the United States was then facing with respect to World War I. The President accepted this offer and requested the Academy to proceed at once to put it into effect.

During the ensuing 2 years the National Research Council was engaged largely in service to the government, acting as the Department of Science and Research Division of the Signal Corps of the Army, and in cooperative relationships with other branches of the government to meet military needs of the country.

With the approach of the termination of the war, the earlier conception for the establishment of the National Research Council was again brought forward with the recognition of the services that an organization such as the Council might continuously render the country during peacetime. This was particularly appreciated in view of the new and important possibilities created by the heavy demands on science and research that had arisen through the exceptional necessities brought about by the war.

Accordingly, President Wilson issued an Executive Order on May 11, 1918, requesting the National Academy of Sciences to perpetuate the National Research Council for further and larger service to science in this country as well as for cooperation with the research services of the military and civil branches of the government (see Appendix B). The general purpose of the Council is to stimulate research in the mathematical, physical, and biological sciences and in the application of these sciences to engineering, agriculture, medicine, and other useful arts, with the object of increasing knowledge, strengthening the national defense, and contributing in other ways to the public welfare.

From the time the National Research Council was organized on a peacetime basis, it included among its several divisions both a Division of Engineering and a Division of Industrial Relations, which was later changed to the Division of Industrial Research and still later to the Division of Research Extension. Early in 1924 the Executive Board of the Council merged these divisions into the Division of Engineering and Industrial Research because of their closely allied activities. In 1966 this division was renamed the Division of Engineering.

The Division of Engineering seeks throughout its organization to take the initiative in delineating and studying new situations arising from changing patterns relating to engineering research, with the following objectives as directional guides: (a) to explore, define, and encourage promising activities in the various engineering disciplines; (b) to explore, define, encourage, and conduct interdisciplinary activities involving engineering or industrial research to a major degree; (c) to promote related activities in education and international cooperation; and (d) to exercise leadership that will stimulate and synthesize thinking among the disciplines of service and engineering, and promote their correlation with economic, political, social, and military factors in support of the general national interest.

Background and Origin of the Advisory Board on Highway Research

Early in 1919 the Division of Engineering of the National Research Council invited several national societies and organizations, including the American Society of Civil Engineers and the Western Society of Engineers, to appoint representatives to the Division. The president of the American Society of Civil Engineers appointed Anson Marston, Dean and Director of the Engineering Department, Iowa State College, Ames; H. H. Porter, President of the American Water Works and Electric Company, New York; and George S. Webster, Director of the Department of Wharves, Docks, and Ferries, Philadelphia. The representative of the Western Society of Engineers was Arthur N. Talbot, past President of the American Society of Civil Engineers and professor of engineering at the University of Illinois.

These civil engineering representatives of the Division of Engineering, with the approval of Division Chairman C. A. Adams, held a conference at the Western Society of Engineers building in Chicago on October 8, 1919, where preliminary steps were taken for the inauguration by the Division of a national program for highway research. Professor Talbot, with the approval of the other civil engineering representatives, had arranged with T. H. MacDonald, Chief of the Bureau of Public Roads, for this highway research conference.

The conference was attended by Professor Talbot and Dean Marston, representing the Division of Engineering of the National Research Council; Mr. MacDonald, A. T. Goldbeck, T. Warren Allen, and J. T. Voshell, representing the Bureau of Public Roads; and Clifford Older of the Illinois Highway Department, representing the Mississippi Valley Conference of State Highway Departments.

Following the Chicago conference, Professor Talbot, Mr. Porter, Mr. Webster, and Dean Marston submitted a report to the chairman of the Division of Engineering stressing the importance and the need for inaugurating a national program of highway research (see Appendix C). Noting that Congress had appropriated $275,000,000 for federal aid in highway improvement, they stated that: "There is urgent need for very extensive immediate scientific highway research to establish the fundamental data needed by highway engineers in designing and constructing the highways which are to be built."

The Division of Engineering at its executive meeting on October 31, 1919, accepted these recommendations and authorized the creation of a Highway Research Committee. The committee consisted of Dean Marston (Chairman), Professor Talbot, Mr. Porter, and Mr. Webster. The purpose of the committee was "to coordinate and assist the highway research work now being conducted by the United States Bureau of Public Roads, by the state highway departments, by manufacturers, research departments, and commercial laboratories."

Mr. MacDonald, who was Chief of the Bureau of Public Roads and had attended the original Chicago conference, was appointed to the Division of Engineering on December 20, 1919, and thus was in a position to maintain active contact with the newly created Highway Research Committee. The Division also created three committees recommended in the Chicago conference report: Committee on Economic Theory of Highway Improvement, with T. R. Agg of Iowa State College as Chairman; Committee on Structural Design of Roads, with A. T. Goldbeck, Engineer of Tests for the Bureau of Public Roads, as Chairman; and Committee on Tests and Properties of Road Materials, with H. S. Mattimore, Engineer of Tests for the Pennsylvania Highway Department, as Chairman. The name of the last committee was changed in 1920 to Committee on Character and Use of Road Materials.

After a few months of operation under this initial committee organization, it appeared that the overall highway research program could be further strengthened, expanded, and crystallized by a conference attended by representatives from all the agencies interested in the general problem of highway improvement. Therefore, on October 26, 1920, C. A. Adams, Chairman of the Division of Engineering, addressed a communication to the governing boards of certain national organizations, federal and state departments, and educational institutions. This communication stressed the rapid growth in both quantity and intensity of highway traffic, and noted that the old methods of road-building failed to meet demands. An excerpt from this communication follows:

> There is great need for such systematic, thorough, comprehensive research as will place our road building on a sound scientific engineering basis, and enable the country to spend effectively the huge sums involved. This need is realized in many quarters and several agencies are at work; but their work is as yet uncoordinated, incomplete, and overlapping.
>
> To meet this need and to conserve the talent as well as the material resources of the nation, the Division of Engineering of the National Research Council undertook to organize a National Highway Research Program, and to secure the cooperation of the numerous existing agencies. Through such cooperation it is hoped to avoid duplication of effort and to bring about the concentration and conservation of the talent of the country.

Dr. Adams included in the communication a proposal to hold a conference on November 11, 1920, to which representatives of all interested organizations would be invited. The proposed General Advisory Board on

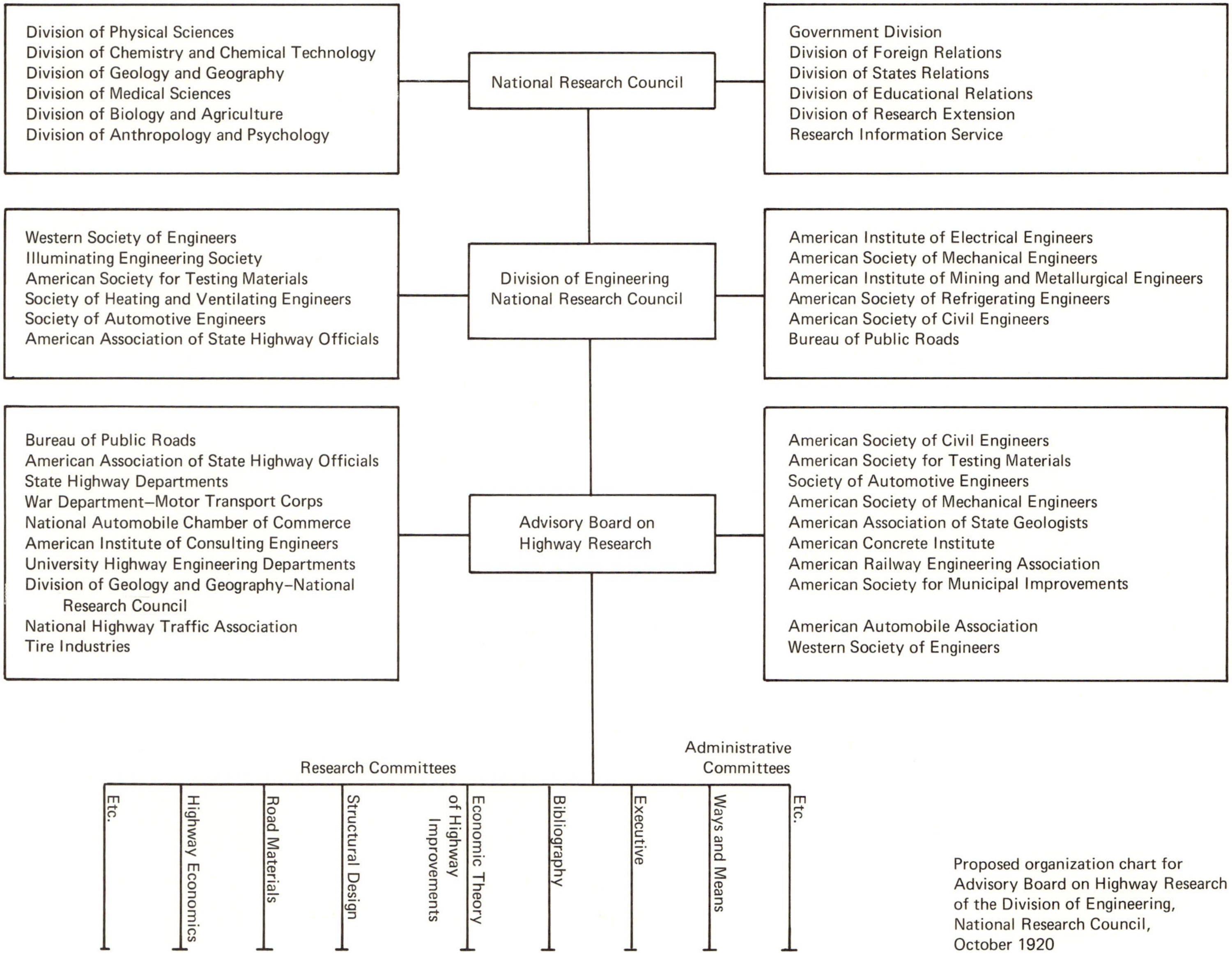

Proposed organization chart for
Advisory Board on Highway Research
of the Division of Engineering,
National Research Council,
October 1920

Highway Research was "to be made up of representatives of interested organizations willing to take an active part, together with some members at large chosen for special fitness. This Board should have a competent full-time, well-paid executive secretary, with a suitable office and office assistants." Its functions would be (a) to comb over the field and select the most important problems; (b) to organize the necessary research committees, including the three existing committees; (c) to deal with ways and means; and (d) in general to serve the purpose implied in the name of the Board.

The communication included a proposed organization chart (see figure) showing the relationships among the National Research Council, the Division of Engineering, the Highway Advisory Board, and the committees of the Board. Those invited were requested to appoint "two delegates to the conference with authorization to take part in the proposed organization on behalf of your Society." The delegates were to be instructed on the following points:

> Will your society be willing to assist in, and support the collection of funds for, the administrative expenses of this movement?
>
> Will your society assist by giving appropriate publicity to the progress of the movement?
>
> In general, is your society sufficiently interested to take an active part, to appoint representatives to the Board and Committees and to give hearty moral support?

The communication concluded:

> The need for the proposed organization is urgent and of interest to every citizen of the U.S. While we regret the necessity for this short notice, we sincerely hope that your society will be able to participate even if only informally.

The Organization of the Advisory Board on Highway Research

The conference at which the National Advisory Board on Highway Research came into existence was held in the Engineering Societies Building in New York City on November 11, 1920. The agenda for the conference had been sent to the invited representatives about a week before the meeting by Professor Adams, Chairman of the Division of Engineering. The conference was called to order by Chairman Adams. The following were present:

Representative	*Address*	*Representing*
C. A. Adams	Washington, D.C.	National Research Council, Division of Engineering

Representative	*Address*	*Representing*
T. R. Agg	Ames, Iowa	American Association of State Highway Officials
David Beecroft	National Highway Traffic Association, 29 West 39th St., New York	Society of Automotive Engineers
Charles J. Bennett	Connecticut State Highway Department, Hartford	American Association of State Highway Officials
Arthur H. Blanchard	University of Michigan, Ann Arbor	National Highway Traffic Association
Roy D. Chapin	Hudson Motor Car Co., Detroit, Michigan	National Automobile Chamber of Commerce
Coker F. Clarkson	29 West 39th St., New York	Society of Automotive Engineers
Henry M. Crane	29 West 39th St., New York	Society of Automotive Engineers
Robert A. Cummings	221 Fourth Ave., Pittsburgh, Penn.	American Society of Civil Engineers
C. D. Curtiss	Washington, D.C.	U.S. Bureau of Public Roads
M. O. Eldridge	Albee Building, Washington, D.C.	American Automobile Association
A. D. Flinn	29 West 39th St., New York	Division of Engineering, National Research Council
A. T. Goldbeck	Washington, D.C.	U.S. Bureau of Public Roads
Prevost Hubbard	25 West 43rd St., New York	American Society for Testing Materials
Hector J. Hughes	223 Pierce Hall, Cambridge, Mass.	Harvard Engineering School
C. D. Jennings	29 West 39th St., New York	*Automotive Industries Magazine*
A. N. Johnson	University of Maryland, College Park	American Society for Testing Materials
Pyke Johnson	Albee Building, Washington, D.C.	National Automobile Chamber of Commerce
H. J. Love	933 Leader Building, Cleveland, Ohio	Committee D-4, American Society for Testing Materials
Thomas H. MacDonald	Washington, D.C.	U.S. Bureau of Public Roads
Anson Marston	Ames, Iowa	Iowa State College
Edward B. Mathews	Johns Hopkins University, Baltimore	Division of Geology and Geography, Association of State Geologists

Representative	*Address*	*Representing*
R. R. McMillan	244 Madison Ave., New York	American Concrete Institute
F. A. Molitor	143 Liberty St., New York	American Institute of Consulting Engineers, Inc.
Daniel E. Moran	55 Liberty St., New York	American Society of Mechanical Engineers
M. C. Morine	International Motor Co., 64th St. and West End Ave., New York	Society of Automotive Engineers
H. de B. Parsons	22 Williams St., New York	American Society of Mechanical Engineers
I. W. Patterson	Room 12, State House, Providence, Rhode Island	State Highway Department of Rhode Island
N. B. Pope	1790 Broadway, New York	Society of Automotive Engineers
H. Hobart Porter	52 Williams St., New York	American Society of Civil Engineers
C. S. Reeve	The Barrett Co., 17 Battery Place, New York	Committee D-4, American Society for Testing Materials
William Sparagen	29 West 39th St., New York	Division of Engineering, National Research Council
Arthur N. Talbot	Urbana, Illinois	Western Society of Engineers
C. J. Tilden	New Haven, Connecticut	Yale University
G. S. Webster	Department of Public Roads, Philadelphia	American Society of Civil Engineers and Division of Engineering, National Research Council
S. Whinery	95 Liberty St., New York	American Institute of Consulting Engineers

Chairman Adams opened the conference with the following remarks:

I welcome you on behalf of the National Research Council, Engineering Division, and include as hosts the Bureau of Public Roads, because of its active and helpful part. Last winter, through Dean Marston, we undertook the organization of a highway research program. That program was started and three committees were appointed. After some experience it has seemed promising of good results to call a conference of all the organizations involved. . . . I take the liberty of suggesting as chairman, Dean Marston, who has been the mainstay in this movement up to date. I also take the liberty of suggesting Mr. Curtiss, of the Bureau of Public Roads, as Secretary of this conference. Mr. Curtiss has been active in the preparation for the conference.

Both of these men were elected to the positions as suggested by Professor Adams. The major preliminary activity of Mr. Curtiss had been the preparation of a draft of bylaws for consideration by the conference representatives. One of Dean Marston's first activities as chairman was the appointment of the following committees, with instructions to report to the conference at the afternoon session:

Committee on Organization and Bylaws—Messrs. Flinn, Cummings, MacDonald, Molitor, Parsons, Chapin, and Patterson.

Committee on Membership and Officers of Advisory Board—Messrs. Talbot, Adams, Bennett, Hughes, Blanchard, Pyke Johnson, Beecroft, and Webster.

Committee on Research and Research Committees—Messrs. Agg, Goldbeck, A. N. Johnson, Hubbard, Tilden, Morine, Reeve, Whinery, and Eldridge.

Mr. MacDonald, Chief of the Bureau of Public Roads, in his address to the conference on "Need for Highway Research," stated:

The problems of reconstruction are of such magnitude and so extensive that they call for the greatest possible encouragement and support on the part of the federal government. Yet in the present situation there is every indication that we are to lose, and lose forever, that one most valuable gift, time, in inaugurating and adequately prosecuting the research and investigational studies incident to a solution of these problems. The highway engineer has for the first time had placed in his hands large sums for highway improvement. Let there be a recognition upon the part of both federal and state legislative bodies that the expenditure of a reasonable proportion of these funds for highway research and experimental studies will be the best investment that can possibly be made. We have placed expenditures for highway improvement at the very top of the list, considering the expenditures for public improvement purposes. We have placed upon the highways nearly 8,000,000 units of traffic, about 12 percent of which are motor trucks. Need there be further argument as to the need for an adequate highway research program on a truly national scale?

Mr. MacDonald was followed by Professor Adams, who discussed the "Need for a National Highway Research Program." He emphasized the importance of cooperation and coordination in research work and stated that the purpose of the Research Council was to bring science and scientific methods to bear on the solution of the practical problems of the day. He also stated that the National Research Council was entering into this cooperative highway research movement in a spirit of enthusiasm and confidence of results and hoped that all the interested organizations would embrace the same spirit.

Chairman Marston then explained the "Plan for Organization of Highway Research Committees." He stated that the committees would fall into two classes: first, those devoting themselves to a single definite research; second, those taking up a group of closely allied studies. The work in general would be handled by numerous committees of the first class. He emphasized the necessity for having these committees chaired by scientists actively engaged in research and staffed with carefully selected active and qualified personnel. In calling attention to the extensive financial support needed for carrying out this program of research, Chairman Marston advanced the suggestion that the various highway commissions and the federal government could devote a certain percentage of the public money appropriated for highways to this research activity. In addition, the commercial organizations could assist by contributing funds, products, or equipment needed for the research projects.

Brief reports on the activities of the three existing highway research committees were made by Professor Agg, Chairman of the Committee on Economic Theory of Highway Improvement; by Mr. Goldbeck, Chairman of the Committee on Structural Design of Roads; and by Chairman Marston, who read the report from the Committee on Properties and Uses of Highway Materials sent in by Committee Chairman H. S. Mattimore, who was unable to attend the conference.

A general discussion was then held covering lines of research work to be taken up, organization of an advisory board, methods of financing a research program, and time, place, and number of meetings. Much interest was manifested in the movement and promises of support and cooperation were given by all organizations represented.

Reports were made at the afternoon session by the chairmen of the three special conference committees: Organization and Bylaws, Membership and Officers of Advisory Board, and Research and Research Committees. Bylaws were adopted and Dean Marston was elected chairman of the permanent National Advisory Board on Highway Research, to serve for one year. Alfred D. Flinn, Vice Chairman of the Division of Engineering, was selected to act as interim director pending the completion of the organization. (He served from November 11, 1920, until June 30, 1921.)

It was agreed to leave to the Division of Engineering the assignment for obtaining permanent representatives and alternates from the societies and organizations represented at the conference and to invite a few other interested national organizations to become members. Included in the latter class were the War Department Motor Transport Corps, the American Society of Municipal Improvements, the Federal Highway Council, and the American Railway Engineering Association.

The primary goal of the conference group was that the Board be so constituted that it would be an effective agency in stimulating and coordinating all activities relating to a national highway research program and in disseminating the findings, and that it become functionally operative as quickly as possible. Toward this latter objective, letters were sent on December 4, 1920, from the Division of Engineering to 19 societies and organizations asking them to appoint a representative and an alternate to the Advisory Board on Highway Research. A progress report on the Advisory Board was sent on January 6, 1921, to the members of the Division of Engineering. At the February 4, 1921, meeting of the Division, held at the Engineers Club in New York City, the following members of the Executive Committee of the Advisory Board on Highway Research were approved:

Chairman: Anson Marston, Dean, Division of Engineering, Iowa State College; Director, American Society of Civil Engineers; Member, Iowa State Highway Commission.

Vice Chairman: Alfred D. Flinn, Member, American Society of Civil Engineers; Vice Chairman, Division of Engineering, National Research Council; Secretary, Engineering Foundation; formerly Deputy Chief Engineer, Catskill Water Supply for New York City.

Member: Thomas H. MacDonald, Chief, Bureau of Public Roads, U.S. Department of Agriculture; Member, Division of Engineering, National Research Council.

Member: George S. Webster, President, American Society of Civil Engineers; Director, Department of Wharves, Docks, and Ferries, Philadelphia; Member, Division of Engineering, National Research Council.

Member: Charles F. Kettering, Past President, Society of Automotive Engineers; President, Dayton Engineering Laboratories, Dayton, Ohio; Member, Division of Engineering, National Research Council.

The Executive Committee made preliminary arrangements early in 1921 on the selection of a director, but progress in obtaining sufficient operational funds was slow, mainly due to the prevailing business depression and the legal limitations of the state governments to enter into cooperative agreements for expenditure of funds on projects outside of their respective states. It was not until July 12, 1921, when the necessary resources were ensured, that the following announcement was released:

Your Executive Committee takes pleasure in announcing that it has engaged as Director, William Kendrick Hatt, Professor of Civil Engineering and Director of

Testing Laboratories, Purdue University. He is a member of the American Society of Civil Engineers and the American Society for Testing Materials. His work as an investigator in engineering and scientific subjects is well known. Director Hatt began duties on July 1. National Research Council has kindly provided an office for him in its building at 1701 Massachusetts Avenue, Washington, and has offered other facilities. Assistance has also generously been proffered by the Bureau of Public Roads and Engineering Foundation. The salary is at the rate of $10,000 a year.

The initial financial support of the Advisory Board for the fiscal year July 1, 1921, to June 30, 1922, was obtained from the following sources:

Department of Agriculture (Bureau of Public Roads)	$12,000
Engineering Foundation	1,000
Connecticut State Highway Department	1,000
National Research Council	500
Total	$14,500

1

2

3

1 The founders of the National Academy of Sciences are portrayed with President Abraham Lincoln in this painting by Albert Herter. They are (from left) Benjamin Peirce, Alexander Dallas Bache, Joseph Henry, Louis Agassiz, President Lincoln, Senator Henry Wilson, Admiral Charles H. Davis, and Benjamin Apthorp Gould.

2 Three of the men who attended the original organizational meeting of the Highway Research Board in 1920 and who were extremely active in HRB throughout the years are pictured together at the 40th Annual Meeting in 1961. They are (from left) C.D. Curtiss, Pyke Johnson, and A.T. Goldbeck. All three served at various times as Chairman of the Executive Committee of the Board.

3 The headquarters of the National Academy of Sciences, said to be one of the most beautiful buildings in the world, was dedicated on April 28, 1924. Located at 2101 Constitution Avenue N.W. in Washington, D.C., it houses the NAS, the National Academy of Engineering, and their operating unit, the National Research Council.

3

A Chronicle of the First 50 years

Ralph Waldo Emerson observed that: "An institution is the lengthened shadow of one man." Emerson was unacquainted with an institution such as the Highway Research Board, for the Board is truly the lengthened shadows of a great many men. It has been said that the idea of its creation was formed in the mind of Thomas H. MacDonald, then Chief of the Bureau of Public Roads, and it is known that Mr. MacDonald nurtured the Board through 25 lean years until it contrived means through its service function to actually become an institution. Anson Marston also was credited with the idea.

Yet from the beginning there was a mingling of ideas, a sharing of effort that, continuing through a half-century, has produced a unique institution combining the best offerings in ideas and actions from preeminent men of every relevant discipline, from every state, and from many countries around the world.

There have been seven directors and 33 chairmen of the Executive Committee since the Board was transformed from a committee in the Division of Engineering of the National Academy of Sciences. Each director has made an enduring contribution of his own. In addition to molding their own concepts into the Board, the directors have, under the guidance of the Executive Committee, assimilated and molded the concepts of an innumerable host of highway administrators, engineers, and researchers into the structure. The Executive Committee, the chairmen and members of the committees, the assistant directors, and the staff specialists have made lasting contributions to the Board and through it to the world.

This chapter attempts to depict the evolution of the Board during the administration of the seven directors and 33 chairmen, citing and assessing developments during each administration.

Alfred D. Flinn, Vice Chairman of the Division of Engineering of the Na- 1920
tional Academy of Sciences, was selected at the organizing conference in New York City on November 11, 1920, to act as interim director of the Board (then called the National Advisory Board on Highway Research), pending the completion of the organization. Anson Marston served as Chairman of the Executive Committee.

The three committees that had been operating under the prior Highway Research Committee created on October 31, 1919, also became part of the Board. At the organizing conference on November 11, 1920, it was recommended that this nucleus of research committees be expanded to a total of five and that two administrative committees be added.

This interim administration was essentially a holding operation until finances could be arranged for the administration of the proposed work program and employment of a permanent director.

The following excerpt is taken from the Annual Report of the National 1921
Academy of Sciences for 1921:

> Since the beginning of the year (1921) the Advisory Board on Highway Research, established through the activities of the Division of Engineering to organize a national program of highway research, has been seeking sufficient funds to employ a director and provide for the necessary office expenses. These funds are now in sight and Professor W. K. Hatt of Purdue University has been engaged to take charge on July 1.

Thus Director Flinn, through his office as Vice Chairman of the Division of Engineering, made his contribution—sufficient funds to begin work. Administrative expenses were estimated to be $20,000 to $25,000 per year. During Mr. Flinn's tenure, the Board's office was in the Engineering Societies Building in New York.

The purpose of the Board as stated in the Bylaws was to "assist in outlining a comprehensive national program of highway research and coordinating activities thereunder; organize committees for specific problems; deal with ways and means; and act in a general advisory capacity." In the Board's evolution this purpose has continued with intensification and expansion of effort. This objective continues. Functions, of course, have been enlarged.

An excerpt from a letter by L. W. Teller, Sr., dated November 11, 1966, describes the first director of the Advisory Board on Highway Research:

> I have quite a vivid recollection of that first Washington meeting, which I attended as a young engineer. There were not many present but Dr. William Kendrick Hatt, who presided, dignified the occasion and impressed me no end by appearing in striped trousers and a cutaway coat.

Having joined the faculty of Purdue University about 1895, Dr. Hatt was Head of the Civil Engineering Department and Director of the Testing Laboratory there when he received appointment as Director of the Board. He had done research of importance in materials and in portland cement concrete and had gained a national reputation. On leave of absence for 2 years from Purdue, Dr. Hatt began with his usual enthusiasm and aplomb to complete the organization of the Board and move into the operational stage.

It is of interest to note that on December 8, 1921, at its Annual Meeting, the American Association of State Highway Officials, then well established, gave official recognition to the Board in the following resolution: "Be it resolved that AASHO welcomes the establishment by the National Research Council of the Advisory Board on Highway Research, and that the Association pledges its support to the Advisory Board in this much-needed effort to stimulate and coordinate highway research."

On July 11, 1921, the Executive Committee authorized the preparation of cooperative agreements with the Bureau of Public Roads, state highway departments, and other suitable organizations for instituting and coordinating a national program for highway research.

On July 27 Dr. Hatt presented a paper at the University of Maryland on "A Program of Highway Research," which was an attempt to chart the field of highway research in its various divisions and to state the important problems. The magazine, *Automotive Industries,* carried an article by Dr. Hatt based on this paper in its August 11, 1921, issue. The article outlined the questions that needed answers in the field of highway research. It is reproduced in Appendix D.

1922 On January 16, 1922, in the Engineering Societies Building in New York, Dr. Hatt staged the First Annual Meeting of the Board. The 30 registrants included 17 members of the Executive Committee. At the Annual Meeting a cogent statement was made: "As a general rule more is learned in research from failures and the reason for failures than is learned from successes."

Cautionary advice was given also in these words: "Coordination does not mean direction." Some time later the word correlation was substituted for coordination because the latter word connoted central administration.

At the meeting the suggestion was made that a broad highway research census be started. A report in October of a census among 132 institutions disclosed 479 highway research projects under way. The estimated annual expenditures were approximately $300,000.

In September Dr. Hatt published a 4-page brochure in the form of a status report encouraging support of the Board's activities.

Dr. Hatt convened the Second Annual Meeting of the Board just 10 months after the first and in the same calendar year. This 1-day meeting was held at the offices of the National Research Council on November 23, 1922.

Three significant statements made by Dr. Hatt at this meeting deserve comment:

"Data that underlie the as-yet undisclosed principles of slab design should be brought together and analyzed to uncover the principles of action." A half-century later, during which monumental research projects have been conducted to this end, there is a great deal more understanding of the tremendously complex system of forces acting on slabs, but additional research is still needed. (Even in 1922 the roadway was considered to be a layered system.)

"Highway engineers and automotive engineers must join forces in coordinated highway research." Nearly 40 years later the grandson of Henry Ford, then President of the Ford Motor Company, visited with the Board's Director Fred Burggraf to discuss the possibilities of closer relationships in respect to highway research. Mr. Ford said, "My grandfather believed that motor vehicle manufacturers should build vehicles and the highway engineer build roads. He would turn over in his grave if he knew I were here today."

"If representation were secured from steam and electric railways the circle of coordination in land transportation would be complete." A half-century later the concept of complete coordination is becoming a reality in the Highway Research Board.

There were other notable events in 1922. At a conference on Tractive Resistance of Motor Vehicles held by the Board on July 14-15, 1922, research roles were outlined for automotive researchers and highway researchers. Efforts in the former area were discontinued. Four decades later, with the concept of a "tire-road system," there began a renewal of joint interest. In particular, with the advent of nonlocking wheels and the role of incipient skid friction assuming importance to both automotive and highway engineers, a much closer relationship in this area of concern may be expected.

1923 A. N. Johnson began a 4-year term as Chairman of the Executive Committee in 1923.

Upon completion of his 2-year leave from Purdue University, Dr. Hatt tendered his resignation at a meeting of the Executive Committee on July 20, 1923. He stayed, however, until December 31, 1923, returning at that time to Purdue.

At the July meeting of the Executive Committee, Dr. Hatt was authorized to secure a technical assistant. On July 30, 1923, Emil Robert Olbrich, a construction engineer for the North Carolina State Highway Department, was appointed to this position.

A bulletin prepared by the Board, "Apparatus Used in Highway Research in U.S.," was published in August 1923 by the National Research Council.

Together, Dr. Hatt and Mr. Olbrich arranged for the Third Annual Meeting, which was held at 1701 Massachusetts Avenue N.W., on November 8-9, 1923, with a registered attendance of 89.

Some of the more significant statements and developments at the meeting were recorded, including this suggestion by Dr. Hatt: "A Highway Research Information Service should be given careful consideration." Other developments included a listing of papers drafted by Director Hatt, general conclusions from the Arlington and Bates Road Tests, and a description of the functions of the research committees.

During this meeting the registrants made an inspection of the new home of the National Academy of Sciences then under construction.

No new committees were added to the existing six during 1923.

Thus the 2½-year administration of Dr. Hatt ended, and Mr. Olbrich carried on for another 3 months until a new director was appointed.

1924 Charles Melville Upham assumed office as successor to Dr. Hatt on April 1, 1924. Mr. Upham was Chief Engineer of the North Carolina State Highway Department at the time of his appointment as Director of the Board. He had previously been associated with the Massachusetts State Highway Department and was Chief Engineer of the Delaware State Highway Department at the time Delaware's Du Pont Boulevard, reputed to be a pioneer among divided highways in the United States, was being built.

After his 4 years as Director of the Board Mr. Upham served as Engineer-director of the American Road Builders' Association until 1951 and then organized a consulting firm. He was a consultant on many highway projects in the United States and to many foreign countries. He was still actively engaged in his consulting work until his death at age 80 in 1966.

Within a month after Mr. Upham assumed office, the new National

Academy of Sciences Building (see page 27) was dedicated (April 28, 1924), and office space (two rooms) was allocated to the Board. At 21st Street and Constitution Avenue (then known as B Street), this edifice, Grecian in exterior style, has been judged to be among the 10 most beautiful in the world. It stands on land purchased with gifts from more than a score of donors and was itself the gift of the Carnegie Corporation of New York. Inscribed in Greek across the front of the building is a statement by Aristotle:

> The search for Truth is in one way hard and in another easy. For it is evident that no one can master it fully nor miss it wholly. But each adds a little to our knowledge of nature, and from all the facts assembled there arises a certain grandeur.

About 2 months after Mr. Upham became Director, his Technical Assistant, Mr. Olbrich, died at age 33. He had served with great promise less than a year.

On June 15 S. S. Steinberg, for many years associated with the University of Maryland, was appointed Assistant to the Director for the summer months and served until September 15.

Professor H. F. Janda, on leave of absence from the University of North Carolina until January 1, 1926, was employed in the fall of 1924 as Assistant Director. He served in this capacity until December 4, 1924, when he became Secretary to the committees for the remainder of the year. Professor Janda later served for many years as Professor of Highway Engineering and City Planning at the University of Wisconsin.

Mr. Upham was notable in public relations and as an organizer. He was a person of great physical vitality and mental ability as well as having an engaging personality. He began immediately to establish avenues for the two-way flow of information between state highway departments and the Board, accomplishing this through contact men (state representatives) appointed to the Board by state highway departments. A 1924 organization chart lists the names of some 35 state representatives, and at least 10 more were appointed later in the year. These contact men have formed an important component of the Board's structure ever since. Mr. Upham also began to establish the same sort of avenues between universities and the Board.

When the Executive Committee met on July 9, 1924, one of its actions was to establish an ad hoc project committee on secondary roads, and authorize the employment of C. A. Hogentogler on August 1. Mr. Hogentogler, an eminent soils engineer, took leave from the Bureau of Public Roads to devote full time to this project.

The Fourth Annual Meeting of the Board was held in the auditorium of the new Academy Building on December 4-5, 1924, with 273 registrants. Director Upham made an interesting comment at this meeting:

A research reporter should learn how to write informational abstracts. Findings will not be applied until read, and they will not be read until they are presented in abstract, and they will not be presented in abstract until the researcher does it—and tells how to use them.

Other statements on record from this meeting indicate the range of interest shown by those in attendance. "We need accurate statistics on road life. . . . Findings should be presented in condensed summary bulletins. . . . The use of horse and wagon for transport of passengers is now almost obsolete. . . . It is time we knew the causes of (traffic) accidents and quit substituting guesses. . . . One-half of 1 percent of improvement funds should be set aside for research."

The First National Conference on Street and Highway Safety was held. Problems claiming attention at this meeting were as follows:—motor vehicle taxes, traffic, traffic capacity, pavement type selection, use of calcium chloride as curing agent, traffic paints for night visibility, soundness tests for aggregates, dust prevention, snow removal and control, sign color, shape, message, and guidance, rolling and air resistance of motor vehicles, highway transportation costs, subgrade studies, sand-clay roads, impact and static load tests, motor truck impact, fatigue of concrete, stresses in road slabs, skew arch investigation, and grading and proportioning of aggregates.

These were problems of 1924. A listing of modern problems a half-century later will include a large number of these "old faces" in new guise.

In its meeting of December 3, the Executive Committee ordered that the name of the organization be changed to Highway Research Board, effective January 1, 1925.

1925 In 1925 several items of importance were recorded.

S. S. Steinberg, Assistant to the Director, was appointed Assistant Director in December and continued in this position into 1927.

As suggested at the Fourth Annual Meeting in 1924, a summary bulletin of research reports was published of the Third and Fourth Annual Meetings. Another bulletin was published of the Fifth Annual Meeting. The last summary bulletin was published in 1926, covering the Sixth Annual Meeting. These bulletins contained the informational abstracts noted by Upham in 1924 as vital links in the process toward use of findings.

The director recommended the inclusion of city contact men in the Board's structure. Little came of this recommendation until 1931. There was also some discussion of the inclusion of industry contact men, but this was deemed inappropriate at the time.

Four sponsored research studies were under way: steel reinforcement in concrete roads; earth road investigations; culvert investigation; and urban highway finance.

At the Fifth Annual Meeting, held in the Academy Building on December 3-4, 1925, there were 293 registrants. As a result of the growth in attendance, the suggestion was made that admission be by ticket and only 300 tickets be issued. Could they have guessed that more than 3,000 would register at the 45th Annual Meeting in 1966! The suggestion was defeated.

Statements of interest recorded in the *Proceedings* of this meeting include:

> The Highway Research Board gives full credit to those agencies or individuals responsible for the particular research. . . . It is necessary that the design engineer know what stresses will be produced by various wheel loads, and that legislators know what loads may be safely allowed on pavements already built.

Later the concept became "one should know the overall consequences of loads and pavement interaction in a framework of overall socioeconomic factors."

Additional problems discussed at this meeting included: tire wear, gas and oil consumption, stress measurements in concrete pavements, mesh reinforcement, crack control, safety of three-lane highways, traffic congestion, maintenance accounting, and roughometer role.

The Westergaard formulas on slab thickness were presented at this meeting.

A reflection of concern at the time was shown in the recommendation of the Executive Committee for the formation of a subcommittee on coefficient of friction between road surfaces and rubber tires, with particular reference to skidding. A commentary on this is manifest in the renewed interest in this phenomenon in the decade of the Sixties.

During 1925 Mr. Upham secured the appointments of 121 contact men from universities. The first foreign contact man was appointed to represent the University of Mexico.

In October 1925 Assistant to the Director Steinberg had this to say:

> A recent census shows 500 highway research projects under way throughout the country. It is the function of the HRB to coordinate these researches, spread as they are over such a wide territory, to prevent duplication of effort by putting the researchers in touch with one another, and finally to make known to each state the findings which may be immediately applied in practice.

Member organizations numbered 24 in 1926. There were 48 state contact men. The number of committees did not increase; however, the Executive Committee authorized the creation of a special committee on the causes and prevention of highway accidents. In 1926 the Second National Conference on Street and Highway Safety was called by Secretary of Commerce Herbert Hoover. Automobile fatalities totaled 25,000. 1926

RESEARCH COMMITTEES

1926

1. COMMITTEE ON ECONOMIC THEORY OF HIGHWAY IMPROVEMENT

Chairman: T. R. Agg

E. L. Clarke
L. E. Conrad
H. C. Dickinson
H. S. Fairbank
T. K. A. Hendrick
H. J. Hughes
Mark L. Ireland
E. W. James
E. H. Lockwood
W. C. McNown
C. M. Manly
O. L. Waller
W. E. Lay

2. COMMITTEE ON STRUCTURAL DESIGN OF ROADS

Chairman: A. T. Goldbeck

Lloyd Aldrich
H. C. Berry
H. E. Breed
Wm. H. Connell
C. N. Conner
R. B. Dayton
G. C. Dillman
F. H. Eno
W. K. Hatt
G. H. Henderson
E. R. Hoffman
C. A. Hogentogler
C. L. McKesson
W. W. Mack
Chas. H. Moorefield
I. B. Mullis
F. T. Sheets
E. B. Smith
C. M. Strahan
H. M. Westergaard

3. COMMITTEE ON CHARACTER AND USE OF ROAD MATERIALS

Chairman: H. S. Mattimore

B. A. Anderton
R. W. Crum
F. C. Lang
C. S. Reeve
H. H. Scofield
M. O. Withey

4. COMMITTEE ON HIGHWAY TRAFFIC ANALYSIS

Chairman: Geo. E. Hamlin

A. H. Blanchard
N. W. Dougherty
J. G. McKay
John H. Mullen
A. N. Johnson

5. COMMITTEE ON HIGHWAY BRIDGES

(Committee not yet organized)

6. COMMITTEE ON HIGHWAY FINANCE

Chairman: H. R. Trumbower

A. J. Brosseau
Frank Page
F. T. Sheets
Jacob Viner

7. COMMITTEE ON MAINTENANCE

Chairman: W. H. Root

H. K. Bishop
W. B. Catchings
G. C. Dillman
J. T. Donaghey
A. H. Hinkle
Wm. A. McLean
W. A. VanDuzer

In 1926 the Highway Research Board joined the Permanent International Association of Road Congresses (PIARC). This affiliation was continued until 1931.

The project on low-cost improved road construction (an aspect of the earth road investigation) was continuing.

Records for the year indicate that the response of state highway departments to solicitations for regular membership fees had not been favorable. Legal provisions generally did not permit such an arrangement.

The development of the portable weighing scale was noted, brought about by "unbelievable loads" on highways.

Some statements of interest included: "The Board, itself, is not a research agency, but serves as a national clearing house. . . . We need proven conclusions stated tersely and with references. . . . Theories should be stated as such. Too much time is taken to read even what is good and too much time wasted on opinion."

Some concepts, needs, and comments of the day included the following: water-bound macadam was disappearing; banked curves and reduction in degree of curves were becoming necessary; portland cement and its reaction to salt was being investigated; carbon-disk strain gages were developed; water-cement ratio concept was developed by D. A. Abrams; need to build pavements 20 ft wide was becoming obvious; belt-line highways for congested areas were proposed; and weighing aggregates for mixtures was developed by Roy W. Crum.

Additional problems discussed in 1926 included alkali reaction, classification of highways, statewide traffic counts, maintenance of concrete pavement, covering disintegrated concrete roads, depletion of gravel surfaces under traffic, and design mix for surface treatments.

The Sixth Annual meeting was held on December 2-3, 1926, in the Academy Building, with 306 registrants in attendance. A single page from the *Proceedings* (see figure) lists the members of all Board committees of that time. By contrast, the Board's 1967 *Yearbook* contained 17 pages of committee members.

1927

The year 1927 marked the last full year of Mr. Upham's services as Director and also the end of the service of Professor Steinberg as Assistant Director. Thomas R. Agg became chairman of the Executive Committee. The *Proceedings of the Seventh Annual Meeting* were edited by Roy W. Crum, the succeeding Director. A happy occasion in 1927 was the employment of a young girl, Ellen Arendes, who rose to secretary for Director Crum and later took charge of the highly complex records for all HRB committee men. Miss Arendes is still performing this important function in 1970.

The functions of the HRB were unchanged since November 11, 1920. Neither were the basic objectives of the founding fathers, so aptly spelled out by one who had a large part in its founding. Thomas H. MacDonald, who told the audience at an Annual Meeting that the purpose of the Board was: (a) to provide a particular and exclusive opportunity for the research worker to express his ideas and ideals to the public and to his fellow workers; (b) to develop the general conception and promote the acceptance of the dignity and improvement of research in the highway and related fields; (c) to lend encouragement and to help sustain enthusiasm in the necessarily long-continued efforts of the individual; (d) to promote friendliness and a more generous appreciation of results secured through efficient and concentrated attention to the development of new knowledge; and (e) to bring about, through mutual confidence and unselfish desires to serve the public interest, a pooling of resources, data, and new knowledge, and the organization and correlation of efforts to make highway research more effective, efficient, and conclusive. "The real objective of research," he said, "is the most economical, in every sense of the word, and the most efficient highway transport service possible, for the use of the public as a whole. The opportunities of research are unlimited."

During the year the proposed committee on bridges, still not established, was discontinued. The Committee on Causes and Prevention of Accidents was established in February.

The studies on low-cost roads and culverts were completed and reported.

All eight committees reported during the Seventh Annual Meeting, held on December 1-2. There were 366 registrants present. Six papers were presented on safety, including one on railroad crossing accidents.

A bibliography on highway safety, 1923-27, was prepared for the Board by the Bureau of Public Roads.

Remarks made at the Seventh Annual Meeting include a comment by Roy W. Crum: "We need treatment of clays to make them serve as satisfactory road surfaces, rather than the detriment they are today."

Other remarks from this meeting reveal the thoughts of the day: "Science needs sound research and humanization. Unless science is helpful to mankind it has no real reason of being. . . . We inhabit the roads. . . . Experiments must be controlled so that results can be interpreted in terms of all real factors of the problem. . . . When we try experiments on a large scale we bring in many factors which lessen control. . . . As variables increase, interpretation of effect of each variable becomes increasingly difficult, if not impossible."

At the Executive Committee meeting of December 9, 1927, Mr. Upham submitted his resignation. Action was deferred until the next meeting, and Mr. Upham continued in office until January 20, 1928.

Thus ended Director Upham's 4 years of administration. There were six functioning project committees when he assumed office; now there were eight. Contact men were established at all highway departments and at some 120 universities, including the first foreign contact man. Five sponsored projects were undertaken by HRB during these 4 years. The Annual Meetings were growing in substance and attendance. The corporate organization now had a good foundation with avenues of communication penetrating every state highway department and leading engineering colleges. Mr. Upham continued to serve HRB as a member of its Executive Committee through 1951.

The following quotation by Lord Balfour is a fitting closure to this period and a timeless challenge to the continuing endeavors of the Board:

> What is discovered at the University of Cambridge, of which I happen to be Chancellor, is equally of use in Paris, Chicago, Japan and at the ends of the earth. It is a gift to mankind.

Each of the first 7 years of the Board has been treated individually here to provide perspective into the problems, concepts, philosophies, and developments—to give a feeling for the situation of a half-century ago. It is remarkable to find so many of the concerns of that generation the concerns of today —just different aspects of age-old concerns expanded in scope and intensified in degree. 1928

In 1928 Frank H. Eno became Chairman of the Executive Committee, and Roy W. Crum began a lifelong term as director of the Board. Except for the beginning years, the 23 years of Roy W. Crum's administration cannot be detailed individually. Every year was fruitful in its immediate and long-time results, but there is space only to record the monumental works of Mr. Crum's administration. As Mr. Crum expressed it in 1929:

> Many of the topics suggested for research are old, some of them more than one generation. Doubtless the explanation is that some words continue to describe briefly the needs under changing conditions for more and better and more precise determinations. However, there appears to me some indication of lack of collection of knowledge gained in days gone by, of its dissemination and of its use by many who might be profiting from it. Strengthening of the Board's endeavors to coordinate and stimulate efforts all over the country would probably help this situation.

Mr. Crum was a graduate of Iowa State College. Later, from 1907-1919, he was a member of the engineering faculty and Associate Professor and Structural Engineer of the Iowa Engineering Experiment Station. In 1919 he was appointed Engineer of Materials and Tests of the Iowa State Highway Commission, where he gained national recognition for outstanding research in materials and highway construction. He was called from that post to direct the HRB.

A friend of his remarked that Mr. Crum was the "idea man" and that his impact on the HRB during his 23 years of directorship did indeed fashion it for its modern-day leadership in transportation research. Mr. Crum saw the desirable catalytic action of highway research in molding the changing scene in a more objective way. He attempted to lead rather than to wait to be pushed into action.

Another friend through the years described Mr. Crum as a quiet, deep thinker, the mentor of his staff. He was diligent in business. Like an editor, Mr. Crum expressed himself in clear, lucid simplicity, getting to the core of problems quickly.

Several eventful occurrences of Mr. Crum's first year should be noted. Two additional sponsored researches were arranged; consideration was given to appointment of industry contact men, but this was not favored; the second research census was undertaken by HRB and AASHO; 335 projects were reported; permission was granted to translate the bulletins on low-cost roads into Spanish and Portuguese; technical papers were included in the Annual Meeting program; and a 400-page bibliography on highway traffic accidents prepared by the Committee on Highway Traffic Accidents was published.

Problems being studied in 1928 included bond financing, personal factors in highway traffic accidents, rigid pavement design, smooth riding surfaces, aerial traffic surveys, and car cemeteries alongside highways.

In his first yearly report to the Executive Committee, Mr. Crum spelled out the functions of the HRB in an interpretative statement:

> In a broad sense, the object of this institution is to make itself helpful in any way possible in the solution of the many problems that confront the builders and the users of the highways.
>
> At present the Board is organized to function along the following lines:
>
> 1. To assemble the results of research work. There exists at the present time a large mass of valuable information, which, on account of lack of publication, has not made its influence properly felt.
> 2. To disseminate useful research information and to aid, wherever possible, in making practical use of the results of research.
> 3. To suggest the carrying on of needed research.
> 4. To arrange for the correlation of research work and the analysis of results by competent authorities, so that highway interests of all kinds may have the benefit of sound principles upon which to base improvements in the construction, maintenance, and use of the highways.
>
> The foregoing statement of objectives applies also to the committees and special investigations which are the operating units of the Board.
>
> We hope as time goes on to broaden the scope and influence of the Board,

and we request the benefit of your suggestions as to how we may best realize the purposes with which we are charged.

To implement these functions in a practical way the director requested $8,000 for a Highway Research Information Service. A funding for this was started in 1930 but, not waiting for the requested funds, Mr. Crum began to issue news letters in May 1928. These were continued through March 1933.

At the Highway Research Board 1929 was an especially fruitful year. The number of committees had increased to 11, and there were many more in prospect. The unwieldiness of the organizational structure for a continuing expansion of committees became apparent. It was also apparent that the subject matter of this unplanned program of the Board was the result of individual interests and, while each was worthy, it was time now to reflect on overall goals and plan for a coordinated framework of activity within a more flexible organizational structure. Two studies were required, a program and an organism, each designed to complement the other and to fulfill the objectives of the Board. 1929

The organic structure was considered by the Executive Committee and a decision made to organize the Board's research activities into six broad areas of concern, grouping related researches categorically within the appropriate broad areas. This philosophy brought about six departments, with ad hoc committees under these created as needed. All existing committees were discharged on December 13, 1929, at the close of the Annual Meeting, and the new organization was announced in February 1930.

The six sections or departments were as follows:

Section	*Chairman*
Administration and Finance	T. H. MacDonald
Transportation	T. R. Agg
Highway Design	A. T. Goldbeck
Materials and Construction	H. S. Mattimore
Maintenance	B. C. Tiney
Traffic (Analysis, Regulation, Safety)	G. E. Hamlin

It is noteworthy that this form of organization endured 40 years, until a wider orientation toward transportation research together with a decade of accelerated activities called for another regrouping and a modification of the organic structure.

To plan a feasible and effective program that would have a concomitant relationship to the structure, a committee on coordination and program was authorized in February 1929, with Mr. Crum as general chairman and with the chairmen of the several existing technical committees as members. The committee outlined 130 broad problems in the framework of the six fields of concern and reported these at the Ninth Annual Meeting on December 12-13, 1929. (See "Report of Committees on Coordination and Program," HRB *Proceedings,* Vol. 9, 1929, pp. 29-53.)

Additional items of interest in 1929 included the creation of a special committee on rail steel reinforcement bars on highway construction on behalf of AASHO. The use of mathematical statistics was noted in the reduction of a mass of data derived in tests.

Employment of Fred Burggraf for special investigations was approved on February 15.

Mr. Crum again pointed out the need for an information service.

About $500,000 was spent on highway research in 1929—half by BPR and the remainder by 24 states.

The Executive Committee decided that: "All matter that is on its face a special pleading or thinly veiled advertisement for some particular product or process shall be eliminated from the proceedings."

HRB had completed a decade of service, providing some substantial contributions from shoestring funding.

1930 In 1930 the departmental organization was made operative. Four of the six departments were to include 28 project committees. The other two departments were in the process of committee formation.

Many of the 1930 concerns are interesting and challenging 40 years later, including: effect of traffic on design, bituminous material for surfacing, maintenance costs, and physics as related to traffic.

At the Tenth Annual Meeting two addresses contained ideas of reference value. One of these was by Albert L. Barrows, Assistant Secretary of the National Research Council:

> In looking back over the course which the Board has traveled, it cannot escape notice that the Board itself has passed through a period of growth by several stages of development, in setting itself up in this field of endeavor, in limiting its field and in defining its problems, in establishing the basis of cooperation with other agencies without which this Board cannot proceed, in determining the kind of work which needs to be done, and, finally, with the aid of many collaborators, in producing a program of research envisaging the matter as a whole, such as was announced and discussed at the last meeting of the Board in a statement in which the present major problems of highway construction are recognized and correlated. It is upon the carrying out of parts of this comprehensive program that many research agencies, allied with the Board, have been bending their energies during

the past year. The results of your work are to be brought forward and exchanged here today and in the meetings of your committees on the other days of this week.

The other address, by H. S. Mattimore, Chairman of the HRB Executive Committee, included the following statement:

> There is another matter that has been touched on a number of times. Is there danger of duplication among the various committees and organizations in the highway field, especially among the national organizations? There is no doubt that in order to avoid useless work it is necessary to coordinate the work of the various national organizations. This is being done. For instance, the American Association of State Highway Officials functions principally through the application of research results to the standardization of methods and specifications. We have close association with them. In fact we have duplication of membership on the various committees and I may say that the American Association of State Highway Officials welcomes the work from this organization. They are in a wonderful position to apply such work.
>
> We are also in close contact with the American Road Builders' Association with its membership and personnel from equipment producers, contractors, city and county executives, and engineers.
>
> We do not wish to have anyone feel there is any danger of duplication of efforts; in fact, the idea of having these three institutions with separate scopes has been working out in a fine manner.

It was recognized from the beginning that it was not feasible for HRB to enter into the field of policy formulation, but rather to provide facts for those who recommend and make policy. In 1950 HRB and AASHO clarified these relations.

It has been said that there was little scientific research in America before 1930. This, however, was not the case at HRB. Mr. Crum understood the ingredients of scientific research and did much to bring the scientific method into highway research. The philosophy of scientific research is not new. It is only in the evolution of the tools of science that we find new avenues of approach, analysis, and interpretation.

Other items of interest in 1930 included the following: nine special HRB research projects were reported as active; the Executive Committee authorized the director to try to get $8,000 to initiate a Highway Research Information Service; and the Executive Committee authorized the Board to conduct special projects when funds are offered and on approval of the Executive Committee.

Also in 1930, the director spelled out the functions of committees. These were (a) statements of problems and subdivisions into fundamental elements

—one of the most important and difficult tasks; (b) outlines of research projects; (c) promotion of necessary research and help in the coordination of activities; (d) correlation of results; and (e) conclusions and recommendations sponsored by HRB based in general on study of adequate facts and substantiated by data and references.

1931 In 1931 Fred Burggraf was designated as Research Engineer. That year the publication, *Highway Research Abstracts,* was begun as a part of the information service of the Board. This activity, begun in May and with assistance from BPR in the beginning, was carried on by Mr. Burggraf personally for many years.

The director was authorized to arrange for contact men in municipal and county engineering organizations; 54 municipal contact men from 27 states were appointed during the year.

Consideration was given to distributing a considerable number of HRB *Proceedings* abroad, in particular to delegates to the Sixth PIARC. The director of HRB was designated as HRB representative to a PIARC Committee.

Approximately $1,000,000 was spent for highway research in 1931, $600,000 of this by BPR.

A report was published on Rail Steel (Part II of *Proceedings*).

Arrangements were made for liaison with the American Road Builders' Association, and steps were taken to establish a formal liaison with the American Association of State Highway Officials. HRB at that time agreed to sponsor the Bartlett Award jointly with these organizations.

A synopsis procedure was initiated for papers in the *Proceedings.*

Some additional problems considered in 1931 included expansion joints, freeze-thaw tests, curing concrete, compacting fills, low-cost highway bridges, driver education, load limitations on highways, and vehicle headlights.

1932 The year 1932 was a year of stabilization. George E. Hamlin became the Chairman of the Executive Committee. No new projects or committees were added that year. Annual Meeting registration was 281.

Mr. Burggraf left HRB to join the Calcium Chloride Association, where he remained for 8 years.

Some problems calling for attention were subgrade research, resurfacing, mudjacking, and shale in aggregate.

Part II of the *Proceedings* of the Twelfth Annual Meeting consisted of the "Report of Investigation of the Use of Calcium Chloride as a Dust Palliative."

In 1933 three new project committees were added: (a) HRB-AASHO Joint Committee on Roadside Development; (b) HRB-Iowa State College Experiment Station Joint Committee on Use of High Elastic Steel as Concrete Reinforcement; and (c) Committee on Evaluation of Highway Systems. 1933

Emerging problems of 1933 included value of time savings, annual cost of highways, economy of highway improvements, coefficient of friction (skidding coefficient), crack control, highway lighting, alcohol and drivers, and rule of road.

Part II of the *Proceedings* of the 13th Annual Meeting contained the report of the Committee on Curing of Concrete Pavement Slabs.

In 1934 Albert T. Goldbeck was named Chairman of the Executive Committee. George A. Rahn, research engineer with the Pennsylvania Department of Highways, wrote Mr. Crum a letter dated January 3, 1934. Excerpts are introduced here because they reveal a significant crystallization of ideas on what correlation means and how it can be practically effectuated. (A decade later the concept was realized.) 1934

I wish to take this opportunity of thanking you for your invitation to attend the dinner at the Cosmos Club in Washington, on December 6, 1933. What interested me particularly was that in the final summation the majority of the ideas advanced centered in your third objective, viz., "To Provide a Clearing House for Highway Research Information," and personally I am of the opinion that the Highway Research Board can do a lot in this direction.

I will continue in the discussion of this third point, which seems to be a centralized point to the effect that the Board act as a correlation agency of research information. Personally I am convinced that a considerable amount of work is being done in the highway field, both in the research and experimental nature, while the results never appear in print. . . .

One of the speakers brought up the proposition of having a traveling representative or representatives of the Highway Research Board in the field at all times. These men would be confined solely to specific problems. This might be possible if funds were available for this purpose. Now in place of such representatives I am going to suggest a representative who would travel continuously, visiting the various members or agencies represented by the Highway Research Board. The duties of this man would be to look over the data obtained through the medium of various investigations or experiments, and request that this be forwarded to the Highway Research Board for file if he thinks such might prove of value for future reference or publication. In this way I believe that a more coordinated effort would result. In other words it might be found that two or more organizations might be working on identical problems, and if it were possible to get these organizations together and compare data, it certainly would lead to more comprehensive study than if these organizations acted individually. . . .

Of course, we all know that finances are the controlling feature in the activities of your organization, and possibly a method of financing as set forth by Mr.

Ables—that is, have the individual states contribute a set sum per annum toward the maintenance of the organization—might be beneficial. Then possibly as a further service, as mentioned by Mr. Mattimore, it might be possible to charge certain fees to non-members of the Board in proportion to the services rendered. I have in mind in this connection the political subdivisions of the various states, viz., counties, townships, cities and boroughs, also any other organization to which data of this nature might prove of value.

Some topics of concern found in the Annual Meeting program included service life of pavements, effect of heavy vehicles on road costs, coefficient of friction of pavement surfaces, nonskid asphalt surfaces, power requirements on grades, computation of stresses in concrete pavements, high elastic steel as reinforcement, effect of calcium chloride on portland cement concrete pavements, motor vehicle braking systems, toll bridge traffic patterns, traffic capacity, highway signs, and highway lighting and highway accidents.

Part II of the *Proceedings* incorporated a symposium on "Research Features of Flexible Pavements."

At this time there was initiated the general policy of charging for copies of the *Proceedings,* which had been distributed free up to this time.

By the end of 1934 the six committees (departments) resulting from the reorganization of 1930, 20 subcommittees, and seven research projects were active. There were now 48 city contact men from 24 states. A new project committee on stabilized roads was established.

One of the increasing concerns in 1934 was that of traffic flow and traffic control in urban areas. Traffic engineers recognized that a traffic pattern for one community could not be transferred to another. A methodology for developing traffic patterns by sampling methods was sought.

Another concern was that of designing adequate coefficients of friction into wearing surfaces. Much theoretical discussion of physics was incorporated in the *Proceedings*. The following stopping distance formula is a good example of this concern:

$$S = \int_{V_0}^{V_1} \frac{vdv}{fg}$$

in which

S = stopping distance;
v = velocity in feet per second;
g = gravity (32.2 ft/sec^2); and
f = coefficient of friction.

In 1935 the Executive Committee approved a joint arrangement with AASHO for the purposes of conducting and maintaining a highway research census. The AASHO Committee on Research Activities would collect the information from the state highway departments and HRB would collect the information from college, commercial, and other research agencies. This information was to be filed in the HRB offices and made accessible to interested persons. Publication of the information would be as agreed upon by the AASHO committee and HRB. The HRB director would act as secretary to the committee. 1935

The Executive Committee also approved an amendment to an action authorizing the Highway Research Board to join with the American Road Builders' Association to form a committee on development of machinery and equipment for use in construction of stabilized roads. This amendment provided for broadening the scope of the committee to include all kinds of highway construction equipment.

A cooperative project with the Bureau of Public Roads for the purpose of research relating to highway safety was approved by the Executive Committee. The Highway Research Board was to conduct research projects approved and funded by BPR.

Some topics of concern in 1935 included annual road costs, air resistance of motor vehicles, roadside developments, load transfer along joints, longitudinal cracking of concrete, vibration in placement of concrete, accelerated acceptance and laboratory tests, distance and time required in passing maneuvers, and soil stabilization.

The Highway Research Board sharpened its structure by organizing the committees or sections into departments. The new organizational chart listed the following departments: Finance and Administration, Highway Transportation Economics, Design, Materials, Construction, Maintenance, Traffic and Safety, and Soils Investigation.

Open technical committee meetings held a day in advance of general sessions of the Annual Meeting were inaugurated. There were 23 project committees within the departmental framework. The Executive Committee approved the employment of a research assistant to devote the major part of his time to the Department of Finance and Administration. Wilfred Owen was appointed to this position and served through November 1938.

The year 1936 saw an increasing expansion in scope of work. At the end of the year there were 39 committees. H. C. Dickinson became Chairman of the Executive Committe. 1936

A research census was begun in cooperation with AASHO. About 1,500 research projects were inventoried. Highway planning surveys had been organized in 40 states, pursuant to the 1934 Act of Congress authorizing their establishment.

Some topics of concern were financial surveys, operating costs of motor vehicles, safe side friction and superelevation, bond stress, durability of concrete, Los Angeles abrasion tests, traffic estimation, organizing for safety, geophysical methods, and soil stabilization for highways.

1937 In 1937 the Executive Committee classified the membership of the HRB and instituted a membership fee.

The $75,000 safety project undertaken in cooperation with the Bureau of Public Roads continued in the driver testing, speed controls, and speed zoning phases.

The Executive Committee authorized prearranged time limits for paper presentations at the Annual Meeting.

The special project on highway lighting submitted a proposal for a program of research.

Part II of the *Proceedings* incorporated 40 papers presented at a 2-day session on soil mechanics and soil stabilization. Other concerns in highway research were uniform highway accounting, motor vehicle operating costs of rural mail carriers, safety and speeds in highway design, sight distance exploration, antiskid properties of road surfaces, curvature and speed, flexible pipe culverts, tire pressure distribution on pavements, highway cross section design, vibration, maintenance costs, one-way streets, lighting requirements, accident-prone drivers, and the highway safety problem.

1938 In 1938 and 1939 Burton W. Marsh was Chairman of the Executive Committee. Mr. Marsh first served HRB as a member of the Committee on Highway Traffic Analysis in 1929, the start of a distinguished career of service to the Board. Still active in 1970 as a member of the Special Committee for Review of Safety Activities of HRB, Mr. Marsh has the longest continuous service for a committee member. He also served on the Executive Committee from 1938 to 1967.

Some new topics of concern as revealed in Annual Meeting papers included speed-change rate of autos, distribution of wheel load through flexible pavement, erosion control, map-cracking of portland cement concrete, frost action in stabilized soils, adhesion of asphalt, warping of concrete pavement slabs, results of highway capacity studies, hill-climbing ability of trucks, alcohol and traffic accidents, and pedestrian accidents.

Member organizations were classified and membership fees initiated.

In 1939, *Highway Research Projects,* a joint census of projects by HRB and AASHO, was published. Eight subject areas paralleling those of HRB departments were covered, with some 1,500 projects listed. 1939

A lecture course in highway economics was approved for a cooperative endeavor by HRB and BPR, to be given at Ames, Iowa, in 1940.

A paper by Fred V. Reagel in the 1939 AASHO *Proceedings* entitled "Is There Duplication in Research?" provided a description of the Research Correlation Service concept.

The Associate affiliation was inaugurated.

Emerging research subjects in 1939 included trends in financial practices, effect of traffic delays on gasoline consumption, guardrails, pavement design for flexible pavements, motor vehicle passing maneuvers, light as it affects highway travel, durability investigation, separation of opposing traffic stream, sign legibility distances, case history by BPR, AAA, and HRB of fatal accidents, soil stresses, and practices in soil stabilization.

The 20th anniversary of the founding of the Board was observed in 1940. At the Annual Meeting there was a registration of 580 persons. There were 287 Associates affiliated with the Board. Fred Burggraf returned to the Board as Research Engineer in 1940, and W. W. Mack became Chairman of the Executive Committee. 1940

The Executive Committee approved a project on the durability of concrete, to be jointly sponsored by ASTM, ACI, and HRB.

The lecture course in Highway Economics was given at Iowa State College on January 8-12, 1940, under joint sponsorship of the college, BPR, and HRB.

The Executive Committee created a special award to give recognition to the authors of annual meeting papers of outstanding merit. This award became known as the Highway Research Board Award.

Highway Research, 1920-1940, was published, with about 1,300 projects included in the census. Some new research topics included in the Annual Meeting program were motor vehicle terminal facilities, highway planning surveys and uses, automatic traffic recorders, design of nonrigid pavements, bearing capacity of pavements, frequency distribution of speeds, stopping distances from high speeds, advisory speeds for curves, location and design of commercial airports, safe approach speeds at intersections, pedestrian walk signals, and physicochemical testing of soils.

During 1941 and the next 4 years the Annual Meetings were scheduled in cities other than Washington, D.C., to avoid the congestion in transportation and hotels during the war years. Meetings were scheduled for the following 1941

1 Hand labor was cheap and plentiful, and equipment was primitive, when this asphalt highway was built near Pleasantville, New Jersey, in 1921.

2 Traffic jams are nothing new, as can be seen in this photograph of sightseers visiting Washington's Hains Point during the Cherry Blossom Festival in 1927.

3 City highways were beginning to show signs of obsolescence even before the postwar boom. This California traffic scene was photographed in 1944.

4 As the automobile ceased to be a luxury and became a necessity, production soared and evolution brought obsolescence to earlier models. Car cemeteries, such as this one in California, began to make an appearance in the Twenties.

5 Concern for proper traffic control was evident even in 1944, when this photograph was made on University Avenue in Berkeley, California. Note the storage bays for traffic making left turns, the lane striping, and the controlled intersection.

1

2

3

4

5

locations: 1941—Johns Hopkins University, Baltimore; 1942—Hotel Statler, St. Louis; 1943—Edgewater Beach Hotel, Chicago; 1944—Cincinnati (cancelled at request of Office of Defense Transportation); and 1945—Skirvin Hotel, Oklahoma City (actually held in January 1946).

Fred Burggraf was appointed Assistant Director on July 1, 1941, continuing a service that was to span in increasing scope nearly 25 years.

A second lecture conference on Highway Economics was held at Texas A&M University at College Station, under the sponsorship of HRB and the university.

Director Crum attended the Pan American Highway Congress in Mexico City in September 1941. At its business meeting following the congress, the Executive Committee authorized Mr. Crum to explore the possibility of arranging translations of pertinent publications into Spanish for distribution in Mexico and the Central and South American countries.

Municipal contacts were discontinued that year.

Emerging topics of concern in 1941 included future development of highway transportation, relation of highway development to the national economy, economics of highway alignment, psychological factors in highway design, accelerated testing of traffic paint, permits for oversize and overweight vehicles, protection at school zones, energy concept of soil-water phenomenon, and cofferdams for bridge piers.

1942 In 1942 the Board issued a series of "Wartime Bulletins." The objective of these bulletins was to disseminate in usable form the best available information on those phases of highway technology in which common practices had not become established or in which practice had to be modified during the war. The name of this series was later changed to "Current Road Problems."

Fred C. Lang became Chairman of the Executive Committee. At its meeting on December 2, the Executive Committee adopted the following motion:

> That the Chairman and Director be required to develop a program for stimulation of research, including the recognition of the value of research by highway administrative and general public bodies, and for the stimulation of more opportunities for research on specific subjects by engineers who are particularly interested.

A supplement to *Highway Research, 1920-1940,* was published. It contained 250 additional projects, some of which had been missed in the previous publication and others that had been started since the previous inventory.

Some of the topics of research included motor vehicle operating costs, road roughness and slipperiness, analysis of wheel load limits as related to design, effect of calcium chloride on dynamic modulus of elasticity, soil-cement design in North Carolina, and present practices in drainage of highways.

In a discussion by the Executive Committee on the role of the Board in education, Mr. Crum said that he had always thought of HRB as an educational institution: "The end product of our work is usable technical information, but it will be of no value to anyone unless it is learned and put to use by the technical men."

In 1943, pursuant to the directive of the Executive Committee in its business meeting of December 1942, a special committee on a program of stimulation of highway research was activated. The committee submitted its report on October 1, 1943. Some excerpts follow: 1943

> Highway research stems from two sources: (1) original ideas of individual scientists and engineers and (2) the need for solution of specific problems that arise in the course of highway practice. . . . The Highway Research Board has always been concerned with both types of research. For the individual workers we have provided a forum and publication medium for the presentation of their material, and for concerted work on particular problems we have provided correlation and direction through committees of experts.
>
> Much highway research work is done without reference to previous work or to what is going on in other places, and it is probable that a great deal of valuable information is never published. A great deal of good could be accomplished by putting out a field agent or agents who would go from state to state and carry firsthand information about what is going on. Along with that, good library research service that could work up reports on the existing information relating to specific questions would be of great help if the states and colleges would make use of it. A traveling representative could do much to promote the use of what is already known.

Included in this report, submitted by F. C. Lang and Roy W. Crum, were the following recommendations:

> In order to present something specific for consideration it is recommended that endeavors be made to expand the work of the Board in the following respects:
>
> 1. By securing funds for scholarships and grants-in-aid.
> 2. By improving *Highway Research Abstracts* and enlarging its distribution.
> 3. By securing additional funds for committee work.
> 4. By providing a correlation service by means of a traveling representative.
> 5. By sponsoring lecture courses and preparing bulletins on utilization of research.

A related action occurred in 1943 in the American Association of State Highway Officials. A special committee, appointed by AASHO and working in cooperation with the Board, prepared a plan for a highway research correlation service entitled "A Suggested Cooperative Research Program." This plan was approved by AASHO by resolution at its 28th Annual Meeting in Chicago on December 3, 1943:

> RESOLVED, that the American Association of State Highway Officials, recognizing the need for an increased program of correlation and utilization of research, directs its Executive Committee to promote such an expanded program as in its judgment is necessary, and copies of these resolutions be sent to the Highway Research Board.

Included in the report submitted by the AASHO special committee was a grouping of the agencies conducting highway research at that time: the Bureau of Public Roads and other federal departments; state, county, and municipal highway departments; manufacturers of highway equipment; and engineering colleges. The report continued:

> There is inadequate correlation of the work of these groups with each other, of that of the individual organizations within a group, and in some cases of the work of persons in the same organization. . . . The work of research committees is essential to bringing to fruition the efforts of individual workers. Few fundamental improvements in highway practice are the results of one man's work. Mostly, research results are crystallized in practice through the joint efforts of many men. . . .

In its report the committee made the following recommendations:

> It is recommended that the American Association of State Highway Officials recognize the need for correlation of research activities through field correlation and committee effort, and urge the Public Roads Administration to cooperate with it in arranging for financing such activities.
>
> It is recommended that the Highway Research Board be requested to act as the agent of the Association of State Highway Officials to administer the funds allotted to this purpose by the Association and to direct and supervise the activities for which provision is made. . . . To accomplish the objectives outlined in the report, therefore, a maximum of six specialists should be employed. Possibly, at the start, the work might be divided among a lesser number but it is felt that the program, if inaugurated, would soon develop to such an extent as to require the full staff.

At the 23rd Annual Meeting some of the topics of concern included postwar planning for highway research, joints and joint spacing, triaxial testing, stresses and displacements in layered systems, culvert design, timber highway bridges, temperature changes in concrete pavements, projective maintenance, origins and destinations of highway traffic, aerial photography and photogrammetry, and model studies of highway subdrainage.

Included in the minutes of the business meeting of the Executive Committee (November 30, 1943) was this statement: "A primary function of the Board is to provide a forum for the presentation of research work."

During 1943 eight "Wartime Bulletins" were published, and three special projects were active. These projects were (a) durability of concrete—a cooperative project by ASTM, ACI, and HRB; (b) structural design of nonrigid pavements—a cooperative project by BPR, HRB, and The Asphalt Institute; and (c) treatment of subgrade soils with calcium chloride—a cooperative project by the Calcium Chloride Association and the Board.

The Hybla Valley Project was initiated as a cooperative endeavor of The Asphalt Institute, the Bureau of Public Roads, and the Highway Research Board (item b above). The site was adjacent to US 1, 4 miles south of Alexandria, Virginia, near Hybla Valley. An oval asphalt test track with 800-ft tangents was built with fill cross section. Work progressed intermittently for a decade (see Chapter 7).

The Department of Traffic and Operations was organized, absorbing the former Department of Traffic.

1944

In 1944 there was no assembled Annual Meeting of the Board. Scheduled for Cincinnati, the meeting was cancelled at the request of the Office of Defense Transportation due to the strains on public transportation. The Executive Committee met, however, at the Academy Building on February 7, 1945, and gave consideration to several important matters. Stanton Walker became the new Chairman of the Executive Committee.

The Department of Finance and Administration was combined with the Department of Transportation Economics to become the Department of Economics, Finance and Administration. Some of the objectives of the reorganized department, expressed by Chairman Herbert S. Fairbank, were to consider development of expressways in cities, value of time savings, intergovernmental relations, long-range financing, and other projects.

Papers submitted and published in the *Proceedings* included the following topics of concern: implementation of postwar programs, airfield pavement evaluation, accelerated traffic tests, plate bearing tests, design standards for interregional highways, runoff computation, influence of curing methods on strength and wear, thermoplastic traffic paints, pumping action phenomena and corrective action, parking requirements, stresses in cuts and embankments, granular stabilization, highway lighting research, salvaging old pavements, advantages and disadvantages of three-lane pavements, and classification of soils.

Excerpts from an attachment to the minutes of the business meeting entitled, "Highway Research Correlation Service" indicate the groundwork being laid for the establishment of this service:

A special committee appointed by the Executive Committee of the American Association of State Highway Officials working in cooperation with the Highway Research Board has prepared a plan for a "Highway Research Correlation Service." Under the proposed plan the Highway Research Board . . . would act as the agent of the Association to organize and administer the Service by means of field engineers, committee work, and publications.

The plan has been approved by the Committee on Research Activities and the Executive Committee of the Association and by the Board. Now, the recent act of Congress providing federal aid funds for postwar highway construction makes it possible to institute this greatly needed activity. . . .

That research is an essential feature of continued development of highway facilities has been recognized by the Congress in the recently enacted postwar Federal Aid Legislation. In the interpretation of the Act by the Public Roads Administration is found the following with respect to research:

"Sec. 8—Planning Surveys

"One and one-half percent of the total amount apportioned to any State under Section 4 of the Act will be reserved for purposes of engineering and economic investigations and research necessary in connection therewith. States will be requested to match in the statutory ratio the federal funds used for these purposes.

"The state highway departments will be advised that the use of small amounts of the apportioned funds for the purpose of contribution to the expense of maintaining a highway research correlation service by the Highway Research Board will be approved as within the purposes of Section 8, providing the federal funds so used are matched with state funds in the statutory ratio."

A great need in this vital activity, and one that heretofore has been inadequately met, is for the correlation and coordination of the works of many agencies.

The engineers, who would each be specialists in some branch or branches of highway technology, would spend a considerable part of their time visiting the state highway departments and other research agencies in order to acquire comprehensive knowledge of current or past research in their several fields, and in order to disseminate first-hand information.

1945 In 1945, with groundwork for the Research Correlation Service (RCS) laid, action began. The following excerpts are from a letter dated June 13, 1945. by Herman A. MacDonald, President of AASHO that year. The letter gives force and effect to the AASHO endorsement of the RCS:

To give effect to a recommendation made last year by a special committee appointed by the Executive Committee of the Association, arrangements have been made to establish, as a function of the Highway Research Board, a "Highway Research Correlation Service."

The plan proposed has been approved by the Committee on Research Activities and by the Executive Committee, and a prospectus describing the manner in which the service will be supplied and its objectives is enclosed herewith.

As you will note, the plan contemplates that financial support for this useful service will be obtained through subscription by the several state highway departments. For the first year the cost of the service has been estimated at $75,000, and this amount has been apportioned among the states in accordance with the original federal-aid formula, thereby establishing equitable subscription rates to be paid by each of the states.

I believe you will agree with me this is a highly meritorious project, and one which should be instituted without delay. A formal request for your subscription, payable July 1, 1945, will be sent to you shortly by the Director of the Highway Research Board.

The year 1945 became a landmark with the inauguration of the Highway Research Correlation Service—a hope of 25 years coming to fulfillment. The director's report acknowledged the inauguration of RCS with the support of the Bureau of Public Roads and the American Association of State Highway Officials, and stated succinctly its role in HRB activities. The District of Columbia, the territories of Hawaii and Puerto Rico, and 41 states jointly subscribed funds for the initial year of operation. Excerpts from the director's report follow:

The recent enlargement of the activities of the Highway Research Board, which we are referring to by the name "Highway Research Correlation Service," is not something new or different from what we have always tried to do. It is rather an enhancement of our stated functions so that the things that need to be done can be done efficiently and adequately.

It has been known for a long time that the primary need is for teamwork by the various agencies interested in the technological development of highway transportation. And, in essence, that is what has been arranged. Where there are so many technical agencies throughout the country interested in highway technology and its development, there should be some central organization that is thoroughly familiar with all the research facilities available.

The states, the government, the colleges, industry, and the research agencies themselves should have better facilities for knowing what others are thinking and doing. I think it is obvious, and has been for the last 25 years, that a central clearinghouse is a vitally important factor in improving highway technology. We have greatly improved our opportunity for service in this respect.

More facilities have been needed for a long time that would provide opportunities for group study and action. We welcome the opportunity to amplify, and improve, and extend committee and department work on technical problems.

The plan of operation of the Research Correlation Service includes provisions for a staff of competent engineers and other technologists who can work in the various areas of interest, travel about the country, and find out the things we need to know about facilities, personnel, and needed information. They will carry information from one state to another, or to other research agencies. In the progress of this work they will gather for dissemination much valuable and usable information. A series of publications called Research Correlation Circulars will be used for transmitting miscellaneous information of all kinds to the subscribers to the service.

Another phase of our activity is further development of committee work. Everyone realizes that, through careful study by committees of people who really know what they are talking about, great progress can be made and without the crystallization afforded the application of new knowledge to construction processes it will be extremely slow, and only through proper support of committee work can it be greatly accelerated.

At this time we recognize a pressing need for the development of a national program of highway research. To that end we need to become acquainted throughout the country with facilities for research, available personnel and the problems with which highway officials are confronted. This work is well under way and substantial progress has been made in the compilation of a list of needed research projects. Much time is being spent at this meeting in discussions of this matter.

The first specialist, as Engineer of Materials and Construction, employed during the latter part of 1945, was, appropriately, George A. Rahn, who had suggested the RCS in his letter of January 3, 1934.

Mr. Burggraf was made Associate Director of the Board in 1945 and, in addition to other functions, began supervising the RCS staff in this new Board activity.

Wainwright Bridges was employed by the Board as Administrative Assistant to the Director, effective September 1, 1945. During World War II he had worked in a similar capacity with the National Highway Traffic Advisory Committee to the War Department and previously had served for 6 years as clerk of the Committee on Roads of the U.S. House of Representatives.

Some topics of research concern at the 26th Annual Meeting, which was held in Oklahoma City in January 1946, were the following: highway development rights; preparing subgrade, subbase, and base for strength; expansion and contraction of concrete pavement; bearing index of soil; capacity of gutter inlets; aerial photography in identifying granular deposits; pavement blowups; temperature variation in concrete pavement; the parking problem; origin and destination surveys; short-count methods in traffic; heavy axle load frequency; subgrade moisture control; and estimating costs of fixed illumination.

1946 In 1946 the 26th Annual Meeting was returned to Washington, D.C., where the meetings have been held ever since. With attendance continuing to increase, the meeting outgrew the available space in the Academy Building and reached into the nearby Red Cross building and the Bureau of Public Roads auditorium (in the North Interior Building). Roger L. Morrison assumed office as Chairman of the Executive Committee.

The Executive Committee in its annual business meeting on December 6 adopted the following resolution:

> The President's Highway Safety Conference, held in Washington, D.C., on May 8, 9, and 10, 1946, recommended that "research into human and physical factors relating to traffic safety be greatly extended and intensified"; and that "agencies in a position to conduct such research should be urged to undertake projects and to contact the Highway Research Board, which is in a position to suggest worthy projects and otherwise to aid as a research 'clearinghouse'."
>
> To give effect to the recommendations of the Safety Conference, three committees have been organized to carry out a coordinated program for highway safety on a nationwide basis—the Federal Committee for Highway Safety, the State Officials Committee, and the National Committee for Traffic Safety.
>
> In view of these progressive actions the Highway Research Board pledges support to the national highway safety program, as a member of the National Committee for Traffic Safety, and offers to cooperate in any way possible with the State Officials Committee and the Federal Committee for Highway Safety.
>
> Furthermore, the Board urges its member organizations and cooperating agencies to encourage and if practicable to conduct research relating to traffic safety, and offers its services and facilities as a clearinghouse for such research activities.

Wainwright Bridges of the HRB staff served on loan as Director of Publications of the President's Highway Safety Conference.

The report of Director Crum regarding the activities of the year incorporated the following statement about the Research Correlation Service:

> The Highway Research Correlation Service was instituted as the result of a study by a special committee of the American Association of State Highway Officials working with a committee of the Highway Research Board. The officers and staff of the Board are now trying to do the things recommended by this special committee. These endeavors do not represent anything new to the basic purposes of the Board. The Correlation Service merely makes it possible to enlarge and make more effective the things we have been doing in a limited way for the past 26 years.
>
> In accepting the subscriptions of the state highway departments to the Research Correlation Service, the Highway Research Board agrees to collect the available information concerning past, current, and proposed research work relating to highways and highway transportation, of the federal government, state highway departments, colleges and universities, and research agencies; study and correlate the information by means of individual and group conferences and committee activities; prepare reports and make recommendations based thereon; and dissemi-

nate such information, reports, and recommendations to the subscribers to the Service and to other highway research agencies.

The plan of organization for the attainment of these objectives, in addition to the Annual Meeting, is in three parts: (1) field representatives; (2) conferences and committee work; (3) dissemination of information.

We have now employed four of the technical specialists recommended by the special committee. These are the engineer of design, engineer of materials and construction, engineer of soils, and engineer of equipment and maintenance. We expect soon to have an engineer of traffic and operations. The assignment of these men is to do everything possible to be helpful to the highway researchers throughout the nation and to aid highway administrators and engineers in the solutions of their problems. In the past year we have visited all of the states at least once; 22 of them twice; 16 three times; and 17 four times. Also 69 engineering colleges have been visited. In this process we have had 895 personal conferences with engineers of state highway departments, 91 with Public Roads Administration engineers, 167 with college people, and 69 with others, mostly in industry. The total is 1,222. We have thus become acquainted with the highway department personnel and have learned a great deal about the problems of the individual states. We have conveyed much information to them orally, by correspondence, and through publications. Now that we are acquainted with the men and the problems we should be able to make ourselves increasingly helpful. Our men by no means know all the answers, but they will have an opportunity to observe all that is being done throughout the nation, and to carry pertinent information from place to place. We hope to complete the technical staff by the addition of a man to cover finance and economics.

Also included in this report by Director Crum was a statement that he had asked the chief highway engineers in all the states to list the problems on which they felt further research was needed. HRB obtained information from 46 states consisting of 810 separate items that provided a source of research consideration for several departments of the Board.

During 1946 the Department of Materials and Construction structured its organization into four major divisions: bituminous, concrete, construction, and general materials.

Some of the research problems were urban inventory control sections, wheel bearing capacity of flexible pavements, use of Hveem stabilometer, stream velocities and scour, drainage maps from aerial photography, live loads transmitted to underground conduits, and aerial strip photography in highway engineering.

Another development was initiated this year that has continued to date. The Board began the series of Annual Reports to its "stockholders." These review the highlights of the year in HRB activities. They have become a rich source of information on research evolution and accomplishments of the Board.

In November of 1946 a college student, E. W. Harris, was employed to help out in the business affairs of the Board. He remained with HRB after he finished his education and over the years advanced to the position of Assistant to the Executive Director in charge of all fiscal management, publication production, and personnel. Also in November, Miss Marilou Kilerlane joined the staff. Later she married staff man Robert Damon. Both are still occupying important positions in the Board in 1970.

Over an 18-month period Messrs. Crum and Burggraf had augmented the professional staff so that in March of 1947 it included: 1947

Name	*Position*	*Employed*
Wainwright Bridges	Administrative Assistant to Director	September 1, 1945
George A. Rahn	Engineer of Materials and Construction	September 15, 1945
A. Walton Johnson	Engineer of Soils and Foundations	January 1, 1946
Franklin N. Wray	Engineer of Design	January 2, 1946
William N. Carey, Jr.	Executive Assistant to Director	January 21, 1946
Elmer M. Ward	Engineer of Equipment and Maintenance	March 18, 1946
M. Earl Campbell	Engineer of Traffic and Operations	March 1, 1947

In 1947 the six departments of the Board had an aggregate of 67 committees. In addition, there were two special committees reporting to the director. A joint committee with AASHO was established for continuing maintenance personnel studies. Committees on roadway pavement markings and on highway channelization were organized.

Arrangements were completed by the Committee on Economics of Motor Vehicle Size and Weight for studies of comparative fuel consumption rates on the Pennsylvania Turnpike and paralleling US 30. Additional topics discussed in 1947 were problems in highway taxation, finding additional revenues for cities, continuously reinforced concrete pavement, effect of air entrainment on concrete, treatment of pumping pavements with cement slurries, effect of centerlines on driver behavior, expressways in traffic diversion and generation, fringe parking, winter braking, clay technology, and geotechnical research.

A vital concern to highway departments was the expressed need for an additional 9,000 highway engineers, and the fact that the enrollment in engineering schools, particularly in civil engineering, was not producing new crops of engineers interested in entering the highway field.

In the Director's Report for 1947, Mr. Crum had some comments about the Research Correlation Service, now concluding its second year:

It is difficult to tell the story of the field service. The accomplishments are necessarily intangible and scattered through several hundred offices. The statistics do not convey much. All of the state highway offices, universities, and industrial laboratories have been visited, not only once but in many cases several times. We are now acquainted with the men in highway technological work and with the problems engaging their attention. We have furnished them with hundreds of items of information and have picked up much material for distribution.

A brief description of what these men do may at least indicate the possibilities. The five engineers now on this work travel about half of their time. On their visits to highway departments and other research agencies they try to confer with all who want to see them, but without wasting the time of busy men in merely polite conversation. After each visit, and before going on to the next port of call, the engineer writes a report on all pertinent subjects discussed and mails it to the office. There it is duplicated and copies distributed to all staff engineers. On his return to the office the engineer goes over the accumulation, not only of his own reports but also those of his colleagues, and notes all matters in his specialty requiring correspondence and transmission of requested information. Taking conscientious care of these details is no small job.

Before starting on his next trip he goes over all of the previous reports of visits to the departments on his schedule in order to be prepared to make his calls of maximum usefulness.

Besides all this, the staff engineers are each assigned to the assistance of the department and committee chairmen concerned with his specialty, and spend much time in keeping track of and helping in the committee work.

Furthermore, the staff engineers have many general assignments, such as collecting material for the highway research census, preparation of bibliographies and project statements on the projects listed in the overall research program, and attendance at highway meetings and conferences.

1948 In 1948 F. V. Reagel began serving as Chairman of the Executive Committee. The Board and AASHO had recognized the need for formalized mechanics for arranging for the conduct of research work by HRB for two or more member states. It appeared that the Board might accelerate its conduct of work for states and possibly engage in work for several states with common concerns. This recognition led to the formulation of a procedure that was approved by AASHO on September 24, 1948.

The Highway Research Board Distinguished Service Award was established to recognize outstanding achievement in the field of highway research. Outstanding achievement consists of distinguished service, production of fundamental research, or administration, promotion, or fostering of outstanding research which in the judgment of the Executive Committee is worthy of the award. (In 1952 the Executive Committee redesignated this award the Roy W. Crum Distinguished Service Award as a memorial to Mr. Crum.)

A lecture conference on the Technology of Low and Medium Cost Roads was held at Salt Lake City on March 1-15 under the aegis of the University of Utah, the State Road Commission of Utah, and the Highway Research Board.

A report was made by the Board of its study of the highway laws, organization, and procedures of the State of Colorado.

Studies of fuel consumption by heavy trucks were made on the Pennsylvania Turnpike and paralleling US 30 by the Committee on the Economics of Motor Vehicle Size and Weight.

The following record from the minutes of the annual business meeting is of interest in that it calls attention to the circumstances in 1948 that figured 20 years later in a reorientation of objectives of the Board:

> Mr. T. H. MacDonald discussed the relation of highway research to research concerning other forms of transportation and raised the question as to whether the Board is giving sufficient attention to cooperation with organizations in other transportation fields. He cited several questions in the general field of transportation affecting highway, railway, and air transport that need to be solved. Among them were questions about location and design of airports with relation to highway facilities. He mentioned particularly the growing movement of heavy freight from the railways to the highways, although it seems more logical for this class of freight to be moved by rail. Since there must be some economic reason for this condition, an impartial research into the underlying factors from the standpoint of the welfare of transportation as a whole should be valuable to all concerned and he recommended that the Board take some steps to look into the matter.

After some general discussion, which was in agreement with the thought that had been advanced, the following motion was presented and adopted: "That the chairman and director be requested to explore the possibility of cooperative research with the railways on problems of mutual interest."

Wainwright Bridges resigned from the HRB staff July 1 to accept an appointment as executive secretary of the State and Local Officials' National Highway Safety Committee.

Discussed in 1948 were such topics as highway needs studies, coordination of highways and city planning, value of travel time, rational design of pavements, hazards of railroad grade crossings, properties of air-entrained concrete, dynamic modulus of elasticity of concrete, welded bridges, reflectorized traffic paints, land-use traffic generation, highway smoke studies, classification and identification of soils, lime-soil mixtures, job evaluation, and night visibility.

1 Detlev W. Bronk, President of the National Academy of Sciences for many years, was one of the first to call attention to transportation problems of urban America. At the HRB Annual Meeting in 1954 he urged the Board to become involved with urban planning, and as a result the Board's Special Committee on Urban Transportation Research (later the Department of Urban Transportation Planning) was formed.

2 Miss Ellen Arendes (left) joined the Highway Research Board in 1927 as secretary to the director, Roy W. Crum, and in 1970 has the longest record of service of any Board employee. Mrs. Robert Damon (right) joined the Board in 1946 as Marilou Kilerlane before marrying staff engineer Damon. Both she and her husband occupy key positions with HRB in 1970.

3 Four of the "giants" in HRB affairs throughout the years were pictured at the 1948 Annual Meeting. They are (from left) C.H. Scholer, Kansas State University; Frederick V. Reagel, Missouri State Highway Department; G. Donald Kennedy, Portland Cement Association; and T.H. MacDonald, U.S. Bureau of Public Roads.

1

2

3

In 1949 the Executive Commitee gave its approval to several ideas. "Procedure for Formulation and Administration of Research Projects Recommended by Committees of the American Association of State Highway Officials To Be Financed Jointly by Two or More States" as approved by the AASHO Executive Committee was printed in *Highway Research Review,* Series 1, No. 2. 1949

The Executive Committee approved a report of the Committee on Extending Correlation Service to Counties. The report contained the following recommendations: (a) that the Board initiate correspondence with states where the county organization situation seemed particularly favorable; (b) that personal contacts be made through the field staff of the Board; and (c) that at the next Annual Meeting one or more sessions be arranged on county road problems.

The formation of an Advisory Committee on the Research Correlation Service was also approved. The committee was to consist of representatives of the highway departments of states, territories, and the District of Columbia and was to replace the present list of contact men with the state highway departments.

In June 1949 the first issue of the *Highway Research Review* series was published. Seven issues were published at irregular intervals as inventories of research in progress throughout the United States. (In September 1962 these items were published in *Highway Research Abstracts* and in July 1963 in the *Highway Research News.* Since September 1965 these items have been published in a series of publications begun by the Board's HRIS, *Highway Research In Progress.*)

The following statement from the annual report of the Department of Economics, Finance and Administration brought into focus the role of the Board with respect to the economics of motor vehicle size and weight:

> The ultimate aim of this study is a determination of the maximum weight of vehicles which will correspond with a minimum cost of highway transportation, costs of vehicle operation and highway provision both considered. This determination is greatly needed as the basis for decisions regarding the weight of vehicles which highways and bridges should be built to accommodate, and also for the eventual corresponding modification of vehicle regulatory laws which for the present must be based upon the capacity of existing roads and bridges.

At a previous meeting of the Highway Research Board Executive Committee (December 10, 1948), particular mention was made of the increasing diversion of heavy commodities from railways to highways, notwithstanding the peculiar fitness of railways to carry this type of commodity. Commenting on the value of impartial research into the underlying factors causing this diversion, T. H. MacDonald suggested that the Board undertake an exploratory project to determine possibilities of research in this regard.

A motion was passed requesting the chairman and director to explore the possibility of cooperative research with the railways on problems of mutual interest. A preliminary exploration had been made during 1949 and reported to the Executive Committee on December 16. After due consideration the following motion was made and accepted: "That it be the opinion and policy of the Board that the study be continued and that the report and the suggestions made at this meeting be referred to the Executive Committee for consideration and action."

Later records do not disclose further activity, but these concerns indicate the growing recognition of the need for research into the interacting forces of the several modes of transportation.

Some emerging topics of concern in 1949 included financing a nationwide highway program, bond issue financing, rational design of bituminous paving mixtures, soniscope testing, fly ash in concrete, glass beads for reflectorizing traffic paint, salvaging old pavement, axle load and gross load trends, accident analysis for program planning, principles and techniques of soil identification, soil temperature studies, and job evaluation.

During 1949 Part II of the "Report of Study of Highway Laws, Organization, and Procedures of Colorado" was completed and sent to the Colorado State Highway Department. Many of the recommendations were accepted and put into practice.

Another milestone in the Board's technological advance was the completion for publication of the *Highway Capacity Manual.* Unfortunately the Board could not find funds for its publication, so arrangements were made for its publication and sale by the Government Printing Office. This publication quickly became very popular, and nearly 30,000 copies were sold in a series of reissues. It was translated into nine languages.

The purposes of the Board were reiterated in 1949 as follows:

> To encourage research and to provide a national clearinghouse and correlation service for research activities and information on highway administration and technology, by means of: (1) a forum for presentation and discussion of research papers and reports; (2) committees to suggest and plan research work and to correlate and evaluate results; (3) dissemination of useful information; and (4) liaison and cooperative services.

In a brochure about the Board Mr. Crum made this statement: "Experience is a great teacher, but by itself it may only deepen a rut."

In 1950 Ralph A. Moyer assumed duties as Chairman of the Executive 1950
Committee. A joint committee was created to outline the most effective cooperative channels for committee activities in AASHO and HRB to obviate duplication of effort and to work within the stipulated policies and objectives of each organization. The committee issued a report on a plan of action for cooperation between the two organizations. A summary of the report is in Chapter 5.

Among other activities of note during 1950, the Board accepted administration of a research project on Wind Tunnel Tests on Bridge Models at the request of AASHO.

Mrs. Marian Strieby, an effective and faithful employee, joined the Board in August. She is still on the staff in 1970 as Administrative Assistant in charge of publication sales and membership records.

Administration of a research project on Intergovernmental Relationships in Highway Affairs was accepted, to be supported by the Bureau of Public Roads, Automotive Safety Foundation, Council of State Governments, American Municipal Association, and National Association of County Officials.

Bulletin 26, *The Truck Weight Problem in Highway Transportation,* was published. In producing this document the correlation staff personally canvassed the state highway departments to obtain factual information on problems in planning, design, construction, maintenance, and operations. This information was supplemented by statistical information compiled by the highway departments and by the Bureau of Public Roads.

In 1950 it was decided to change the date of the Annual Meeting from December, when it had traditionally been held since the mid-Twenties, to January in order to avoid conflict every other year with the AASHO Annual Meeting. Thus there was a 13-month interval between Annual Meetings, and none was held in 1950.

The 30th Anniversary of the Board was celebrated at a special session of the 1951
30th Annual Meeting in January 1951. The following note was inscribed on the program:

"Of those mentioned in the earliest records, Thomas H. MacDonald, C. D. Curtiss, A. T. Goldbeck, H. S. Fairbank, C. M. Upham, Pyke Johnson, W. H. Root, and C. N. Conner are still active in the Highway Research Board." Mr. Curtiss had acted as secretary at the organizational meeting on November 11, 1920. Highlighting the program of the special session were addresses by Detlev W. Bronk, President of the National Academy of Sciences, and Thomas H. MacDonald, U.S. Commissioner of Public Roads. The following excerpts from a report on these addresses reveal the thinking of an eminent scientist and an eminent administrator:

The necessity for the continuation of basic research during the present emergency and for speed in utilizing research results was forcefully presented by Dr. Bronk. Commissioner MacDonald described the formation of the Board and reviewed the achievements in highway research during the past 30 years.

In mentioning the cooperative research activities of the Highway Research Board, Commissioner MacDonald emphasized the importance of this type of effort. It is clear that the resources of the federal government, the state highway departments, and industry should be combined for an attack upon the many far-reaching problems of mutual and general interest if highway problems are to be solved quickly and efficiently.

About the 30th Annual Meeting itself, Director Crum had these comments:

The Annual Meeting has grown into a large and far-reaching affair. Basically it was set up 30 years ago to be a forum for the highway research workers and it has been so operated ever since. We endeavor to make a place on this rostrum for all of the research reports that come to us each year. The Correlation Service has greatly stimulated this phase of our work. It has brought us from a meeting of 11 sessions with 41 reports in 1945 to the present meeting (5 years later) at which it has been necessary to arrange 31 separate sessions to provide for the 159 papers and the additional discussion periods needed. This Annual Meeting has been attended by 860 representatives of federal agencies, states, colleges and universities, industries, and others interested in the technology of highway transportation. The papers and reports presented emanated from 75 different organizations.

Also from the director's report is this terse statement: "The test [Road Test One–MD] begun in June 1950 of the effects of different truck axle loads upon a concrete pavement by means of controlled traffic is nearing completion. A progress report giving the results to date has been made at this Annual Meeting."

This signaled the start of the great series of field tests to study pavement versus loading administered by the Board in the 1950's. Asriel Taragin, on loan to the Highway Research Board from the Bureau of Public Roads, served as Project Director on the Maryland Road Test.

At the 30th Annual Meeting some of the emerging concerns were continuous reinforcement in concrete pavement, saw cutting of contraction joints, high-strength bolts, heating bridge decks and concrete pavements, calcium chloride as a strength accelerator, speeds on rural highways, vibrations produced in structures by heavy vehicles, lime-fly ash in highway construction, and stabilizing swamps.

The 30th Annual Meeting was the last Annual Meeting conducted by Mr. Crum. During the week prior to May 6 he went to Columbus, Ohio, to preside at a committee meeting. H. S. Fairbank was there and said of Mr. Crum's performance that he had never done a finer job. Upon Mr. Crum's return home on May 13 he was stricken with a severe heart attack and died within a few hours.

At the 31st Annual Meeting, the Distinguished Service Award of the Highway Research Board and the George S. Bartlett Award were posthumously awarded to Mr. Crum. These awards were received on behalf of Mr. Crum's family by Fred Burggraf. The following statement was made by Thomas H. MacDonald in his eulogy to Roy W. Crum:

> Roy Winchester Crum left three heritages:
>
> An organization grown from nine committees and 81 members to one of 80 committees and 758 members.
>
> A 5-foot shelf of the finest in highway research publications.
>
> The concept that highway research is a continuing, unfolding process and that new men year by year should be encouraged to share in such activity.

The following statement is incorporated in the dedication memorial in the *Proceedings* of the 31st Annual Meeting:

> To a far greater extent than is generally realized, American highways have provided increasingly improved services because of knowledge gleaned from researches encouraged and on occasion directed by the Highway Research Board. Unobtrusively, indefatigably at work behind the scenes were ever the guiding hands, heart, and mind of our respected Director. Millions of highway dollars have thus been conserved; but no monetary estimate can fairly evaluate his outstanding professional contributions. . . . Inspiration to continuing research advancement comes from the fine human qualities of this man. His integrity, friendliness, modesty, open-mindedness, industry, fairness, determination and resourcefulness have won universal respect and have endeared him to all—young and old—who have in one way or another been associated with him.

Thus 1951 marked the beginning of a new administration and an acceleration of activity. If it was vexing to select the most significant events among the activities of the previous administration to include in the Board's evolution and accomplishments, it becomes increasingly difficult to telescope the activities of the successive years into a condensed version of accomplishments.

Mr. Crum was succeeded by Associate Director Fred Burggraf. Named Acting Director on June 12, 1951, he was appointed Director on November 1.

Mr. Burggraf had attended George Washington University in Washington, and in 1916 joined the National Bureau of Standards as a member of its Structural Materials Division. From 1920 to 1928 he was a Research Engineer with the Illinois Division of Highways and participated in the famous Bates Road Test, the first large-scale test of pavement behavior under controlled traffic.

Mr. Burggraf first joined the Highway Research Board in 1928 as a Research Engineer in charge of special projects. Four years later he left the Board to become a staff member of the Calcium Chloride Association. During the next 8 years he supervised, in addition to other duties, university and Bureau of Standards fellowships for the Association. During this time he also developed the Burggraf Shear Machine for field testing the shear strength of the individual layers of flexible pavements, as well as field procedure for controlling the use of calcium chloride in concrete to balance high early strength with ease of surface finishing. He returned to the Highway Research Board in 1940 as Assistant Director. He was named Associate Director in 1945 and was made Director in 1951.

In 1951 the field work on the Maryland Road Test was completed and work on the WASHO (Western Association of State Highway Officials) Road Test was begun under W. N. Carey, Jr., who was assigned as Project Engineer (see Chapter 7). Among the special projects being conducted by the Board in 1951 were Effect of Wind Stresses on Bridges, a cooperative project initiated by AASHO and sponsored by a group of states, and Relationship of Parking to Business, a 2-year project directed and administered by the Board with a grant of $70,000 made available by the automotive and petroleum industries through the Automotive Safety Foundation. The actual investigations were delegated to the business schools of several selected universities.

The *Annual Report* of the Board for 1951 included "Trends in Research," which reviewed and interpreted the nationwide trends in highway research. This was reinstituted in 1954 as "The Year in Review" and has been a part of the *Annual Report* since that time. It provides a good general source for highway research history.

A trend in the early Fifties was the inauguration of high-speed electronic computers in highway research, engineering, and accounting procedures. The first digital computer, ENIAC, was switched on at the U.S. Army Proving Grounds at Aberdeen, Maryland, in 1946. The electronic computer soon reached into the highway field and made its tremendous impact in acceleration of nearly every reiterative computation and other computations that would have required too many unavailable man-hours.

1952 In 1952 R. H. Baldock became Chairman of the Executive Committee.

The final report of the Maryland Road Test was published, and the demand was so great that 9,000 copies were printed during the year. The

Board also completed and published its study of state highway administrative bodies, as well as Special Report 5, *Channelization*—one of its most popular publications.

Some of the problems considered in 1952 included capital investments in highways, incremental tax on motor vehicles, composite construction in highway bridge designs, prestressed concrete in highway bridges and pavements, ultrasonic pulse technique in testing, bridge maintenance practice, effect of shoulders on speed and placement, speed and volume related to motor vehicle accidents, braking distance from high speeds, and composition and engineering properties of soils.

In September, Kenneth G. McWane joined the staff as Engineer of Traffic and Operations.

Also in 1952 an experimental skid-test road was built in California.

Looking at the urban scene, the Detroit Area Traffic Study was the first comprehensive metropolitan area transportation study and a pioneer in the development of traffic assignment techniques and land use traffic generation modeling.

1953

One of the most significant events in 1953 was the retirement of Thomas H. MacDonald on March 31 after 34 years as Chief of the Bureau of Public Roads. During each of those years he had been a staunch supporter of the Board, and his impact on the role of the Board has endured. Upon retirement he went to Texas A&M University where he organized and directed the Texas Transportation Institute until his death in April 1957.

Mrs. Beatrice G. Crofoot joined HRB in November 1953 as a clerk-typist. Since then she has advanced to the position of Production Manager of the Editorial and Production Office, where she is responsible for the printing and publication schedule of HRB publications.

Special Report 15, *Highway Research Organizations,* was published by the Board. Written by M. Earl Campbell, its purpose was to describe existing organizational patterns and activities of the several state highway departments in their research endeavors.

The report of the Committee on Roadside Development stated that: "The aim of the committee is to search for, assemble, and evaluate roadside development methods and practices throughout the country and to disseminate to all concerned. . . ." By assembling packages of published material from various states and sending to other states, an effective clearinghouse was maintained by this committee.

Problems being discussed in 1953 included an urban congestion index, the relationship between auto and highway, statistical methods in highway research, accident-exposure index, dynamics of guardrail systems, performance of bituminous mix design, control of reflection cracking in overlays, gap acceptance in the traffic stream, shear stress in soil masses, and a system of soil classification.

Another event was the establishment of the Committee on Highway Laws in the Department of Economics, Finance and Administration. It was established by the Board in response to a 1951 resolution of the Executive Committee of the American Association of State Highway Officials. The objectives of the committee are (a) to assemble and analyze state constitutions and highway statutes both state and local in character in terms of their functional parts, as they pertain to system classification, highway design, control of access, traffic engineering, land acquisition, construction, maintenance, and other matters relating to roadbuilding activity; and (b) with the background of fact so derived, to isolate the essential principles and elements that constitute adequate highway laws in each functional field delineated. This committee became the Special Committee on Highway Law in 1960 and the Department of Legal Studies in January 1964.

1954 Walter H. Root was elected Chairman of the Executive Committee. He served as Chairman only a short time before he died of a heart attack on April 27. G. Donald Kennedy succeeded Mr. Root and served through 1955.

In 1954 the WASHO Road Test was completed (see Chapter 7).

Dr. Detlev W. Bronk, President of the National Academy of Sciences, in an address to the delegates at the Annual Meeting urged HRB to concern itself with the problems of urban America. In response to Dr. Bronk and others, an ad hoc committee on urbanization research was created in April. This committee eventually became the Special Committee on Urban Transportation Research and then the Department of Urban Transportation Planning on June 20, 1963.

Elmer Ward assumed the post of Assistant Director after serving 8 years as Engineer of Materials and Construction, 6 of these years as Engineer of Maintenance and Equipment. Mr. Ward obtained his BSCE degree from Iowa State College in 1927; while there he had studied under Mr. Crum. He worked for the Iowa State Highway Commission in the Materials and Tests Department until 1941, then spent 5 years on research and development of road-building equipment in the Roads and Airfields Branch of the U.S. Army Corps of Engineers. He began work for the Board at the invitation of his former teacher, Mr. Crum.

Mr. Ward served as Assistant Director until his death in August 1961. He was given, posthumously, the Roy W. Crum Distinguished Service Award, and his memorial noted that he had served as Assistant Director with "loyalty and devotion to the work and principles of the Highway Research Board."

Subjects of concern in 1954 included sociological research in urban passenger transportation, design of toll plazas, bridge pier scour, pavement slipperiness, traffic paint performance tests, nondestructive testing of concrete pavements, rebound hammer tests, widening and resurfacing with concrete, quality of traffic flow, assignment of traffic to toll facilities, effects of repeated loading on clay, and chemical soil stabilization.

The Chicago Area Transportation Study was initiated, building on the earlier Detroit study, and became the first of the new generation of transporation planning processes.

1955

A number of noteworthy events occurred in 1955. On February 22 the Board agreed to undertake the administration and direction of the AASHO Road Test. The preliminary details had been in the hands of a Working Committee of the Highway Transport Committee of AASHO for several years. With the WASHO Road Test completed the Board was ready to turn its attention to this new undertaking, which was to cost $27 million and last until 1962. W. N. Carey, Jr., who had been a member of the Working Committee, was assigned to supervise the research, and W. B. McKendrick, Jr., was assigned as Project Director of the large project. A project Advisory Committee was established under the chairmanship of K. B. Woods. Paul E. Irick joined the Road Test staff in December.

In January Ray E. Bollen joined the staff as Engineer of Materials and Construction.

The Highway Laws Project, inaugurated in 1955, drafted publications covering 28 areas in highway law before it ended in 1960. It was initially sponsored by BPR and ASF and later was funded by the state highway departments.

With the accelerated pace in highway construction, together with the difficulty of recruiting and retaining highway engineers, the staff of the Board canvassed the various divisions of the state highway departments to find out what time-saving methods were employed in each. This canvass was published in a series of chapters (division by division) during the year.

There were 1,520 people in attendance at the Annual Meeting. Some of the problems discussed were the investment analysis approach to estimating highway needs, benefit-cost ratio applied to highways, lives of highway surfaces, reinforced concrete pipe under earth fills, use of statistical methods in reducing AASHO Road Test data, radioactive isotopes in determination of asphalt content, air meter for air content of hardened concrete, highway accidents related to shoulder width, use of electronic computers in traffic problems, and soils exploration and design considerations.

1956 Kenneth B. Woods became Chairman of the Executive Committee in 1956.

The report of the President's Advisory Committee on the National Highway Program (the Clay Committee) stated the need for greatly increased interstate highway modernization. This report set the stage for Congressional action, which culminated in the Federal-Aid Highway Act of 1956. This Act was the most important legislation on highway research and construction to date. For the first time, legislation sought to equate the motor vehicle with the road it traveled on. This Act also had as one of its aims the completion of the Interstate System.

By 1956 the demand had so increased for lecture and committee rooms that the Annual Meeting was held at the Sheraton-Park Hotel. It has remained there since, and after a few years also expanded into the nearby Shoreham Hotel. In 1956 there were 1,700 people in attendance at the Annual Meeting. Problems discussed there included urban transportation studies, incremental cost studies, determining bridge tolls, future traffic predictors, moisture movement through soils, and evaluation of the structural adequacy of highway bridges.

Herbert P. Orland joined the Board as Editor of HRB publications, and has since become Editor of NCHRP publications.

The construction of the AASHO test road began on August 7.

The RCS staff was moved to the DuPont Circle Building in August.

On August 31 Joseph L. Intermaggio was named Project Director for Research under the newly established Special Committee on Urban Research, which was to include the clearinghouse function by preparing abstracts of urban research. The initial funding was by the Automotive Safety Foundation. (Mr. Intermaggio continued through 1957.)

An inventory of highway engineering manpower and needs was completed by the staff.

Chairman K. B. Woods initiated functional standing subcommittees of the Executive Committee to particularize effort and ensure effective work (see structural description in Chapter 4).

1957 Rex M. Whitton served a one-year term as Chairman of the Executive Committee in 1957.

Pursuant to a request from AASHO in November 1956, the Board held a conference in March 1957 devoted to the economic impact of highway improvement.

In 1957 the registration at the Annual Meeting climbed above 2,000 for the first time, reaching 2,025. Thomas H. MacDonald died in April 1957, and the *Proceedings* of the 36th Annual Meeting were dedicated to him. The following excerpt is taken from the dedication:

When he arrived in the nation's capital in 1919, as head of the Bureau of Public Roads, Mr. MacDonald almost at once began to explore ways and means of expanding highway research, for which $300,000 a year was then available. He was determined to enlarge its scope. He spoke in numerous conferences of the need to coordinate the work of individual researchers in institutions scattered throughout the country. And he emphasized strongly the importance of making the results of fact-finding and analysis widely available to all interested groups and to the public.

In a real sense, the whole work of the Board, since its establishment in 1920 as the Advisory Board on Highway Research, has been a reflection of this man's vision and his unswerving loyalty. He helped to create the organization; he sustained it during those difficult pioneering years when the vital role of research in highway transport was not yet widely recognized; and from his philosophy and his magnificent achievements in this field will come the inspiration and guidance for much more constructive service to the nation and the world in future generations.

Some problems of concern in 1957 were operations and systems research, research related to design of turnpikes, radiographic inspection of welded highway bridges, traffic operations in highway tunnels, electronic control of motor vehicles, foamed asphalt in soil stabilization, and soil wave propagation.

Charles H. Scholer was elected Chairman of the Executive Committee to serve in 1958. The first session on urban research was held at the 37th Annual Meeting. This was particularly timely since at the same meeting, in his introductory remarks. Academy President Detlev Bronk asked questions that were certainly prophetic as one looks back from 1970. He said: 1958

What are the new relations of city to country that are evolving out of the great highways which you are building? What, as I have asked of you on previous occasions, is the future of our cities? What is to be the pattern of urban living, which is certainly the pattern of the future in a world growing ever larger? What is the future of transportation by railroads and by airways, which you are affecting? You are helping to solve some of the transportation problems, but what is going to be the future pattern, which is going to affect the lives of all of us, and all of our descendants? What is the sense and the need of the wasteful space of cars grown ever larger for the transportation of men and women who are no larger than they were before? At a time when we are berating those who have not done as much as we think they should do, what is our capacity to spend ever-larger effort and to consume more raw materials in the building of these ever-larger cars, which make your problems ever greater? What is the ultimate consequence of the destruction of millions of acres of arable lands, and how can we determine the proper course of highways without undue regard for the selfish reactionary interests of certain peoples and certain groups? Should we destroy natural beauty by permitting those with selfish interests to contaminate a scenery which has been made great by nature and is too often squandered by selfish men and women?

Other topics of concern discussed at the Annual Meeting included skid prevention, organizing for continuing work in metropolitan areas, frequencies of various levels of stress in highway bridges, pavement profile surveys related to performance, electronic instruments in geodetic surveys, expanded clay aggregates in bituminous construction, travel time as a measure of congestion and level of service, and shoulder occupancy on rural highways.

In August 1958 the administrative staff of the Board moved from the Academy Building and the RCS staff moved from the DuPoint Circle Building to 1707 H Street N. W., thereby reuniting the entire Board and expediting its functioning.

Some 30 state highway departments had electronic computers.

In the same year the Sagamore Conference, cosponsored by the Special Committee on Urban Research and Syracuse University, was held in October. This conference revealed the necessity for a common understanding of the highway problem in urban areas by the various disciplines concerned.

1959 Harmer E. Davis became Chairman of the Executive Committee for 1959.

The Board's publications comprised a 10-foot bookshelf and were rapidly expanding as reports from accelerating research projects were published. Problems from the developing Interstate System were bringing an upsurge in research studies on all fronts.

The Board published a report in 1959 that was a tremendous stimulus toward making a real dream that the Board had cherished since 1920—a national program of highway research. Special Report 55, *Highway Research in the United States: Needs, Expenditures and Applications,* a cooperative endeavor by the staff of the Board under a Committee chaired by E. H. Holmes, was published. The report made crystal clear that, while problems of research caliber were increasing, the solutions were not keeping pace. Only 2 mills were being spent out of each highway dollar on research, and applications of the research results woefully lagged. This report accentuated the need, listing 101 problems that highway departments supplied, and showed clearly the inadequacy in funding. It also gave 348 citations of findings put to use and the resultant benefits. And the report noted 19 problem areas needing urgent research attention. Although it took nearly 3 years of planning, it was on the basis of this report that the National Cooperative Highway Research Program (NCHRP) was established in 1962.

The Committee on Public Dissemination of Research Findings was established and began to function during 1959.

The Board, acting jointly with the American Society of Civil Engineers and operating with a grant from the National Science Foundation, arranged for exchange visits with soils engineers from the Soviet Union in June and September.

The AASHO Road Test was continuing with the cost estimate now placed at about $27 million.

The Department of Economics, Finance and Administration conducted a conference in September on Economic Analysis in Highway Programming, Location, and Design.

At the Annual Meeting there were 2,610 registrants, 222 papers presented, and 82 committee meetings. Some topics discussed were the status of electronic computation in highway engineering, roadside design for safety, designing for high skid resistance, vehicle design and performance characteristics, human factors in passenger car design, terrain appraisal methods, and proof-rolling of subgrades.

In 1960 Pyke Johnson served as Chairman of the Executive Committee. 1960

Traffic at the AASHO Road Test completed its running on November 30, 1960, with more than a million load applications. Correlative tests continued, but the staff had begun work on analysis of data and drafting of final reports.

The Highway Laws Project ended on May 1, and 28 areas of highway law had been explored by a group of attorneys employed by the Board for this work. Their work was published in the Special Report series. The Committee on Highway Laws was made a Special Committee of the Board in June.

The Special Committee on Urban Research published a circular in May entitled *Suggested Subjects for Urban Research.* These suggestions became the basis for several projects undertaken by NCHRP in the next few years. Other departments have also drafted circulars on their research needs.

Earle W. Jackson joined the staff in July as Assistant Editor. He was to become Senior Editor of HRB publications prior to his resignation in 1969.

The National Academy of Sciences Woods Hole Conference on Transportation (all modes) was staged in August. HRB furnished considerable material on highway transportation and Messrs. Burggraf and Carey of the staff participated in the conference. Again the desirability of research at the interfaces of the several modes was seen, and a broad transportation capability within NRC was recommended. Proceedings of the conference were published by the National Academy of Sciences.

In September the Department of Economics, Finance and Administration sponsored a conference on Formulating Highway Construction Programs.

A phenomenal growth in activities had taken place during the Fifties. Socioeconomic research was coming to the fore. Studies of interrelationships among modes of transportation were gaining momentum. Growth in Annual Meeting attendance and publications had become astronomical. The staff work had tripled since the inception of the correlation service. Concern was being expressed regarding the shape of things to come and how to meet the demands.

There were now six departments, 10 special committees reporting to the director, and 84 project committees. Over 1,000 men filled 1,683 committee assignments.

1961 In 1961 W. A. Bugge was elected Chairman of the Executive Committee.

One of the ways in which the Board celebrated its 40th anniversary was with the publication of Special Report 63, *In Commemoration of the 40th Annual Meeting*. This publication contained papers presented at the anniversary celebration during January 9-13, 1961, including a paper by Director Burggraf entitled "Historical Highlights of the Highway Research Board."

Registration at the 40th Annual Meeting was 2,403. Some topics of concern discussed were the role of planning in highway administration, planning and measuring highway progress, guardrail research, kinetic behavior of roads, alkali reactivity of carbonate rocks, epoxy resins in concrete repair, new highway capacity concepts, and soil compaction and consolidation.

The year 1961 brought into clearer focus the need for additional expansion of the activities of the Board to cope with growing demands. The five RCS engineers, all about the same age, were approaching retirement. In 4 years these men would begin to leave. Information requests were accelerating, which emphasized the need for automated processing of such requests. Secretarial help was inadequate to cope with the problem.

Thus a 10-point program of expansion needs was outlined, including assistants for the RCS men. A request by the Executive Committee for a worldwide research census was made. Another $100,000 annually was needed for funding. The Executive Committee decided to create a new category of membership for industry because nearly one-third of the committee membership and registration at Annual Meetings consisted of highway industry representatives. The membership fee was set at $1,000.

To acquaint industry with the potential values of the Board in its information and correlation service, an "Industry Dinner" was held on September 26 for 100 leaders in the highway industry. Some 25 industries affiliated with the Board as a result of the invitation.

Assistant Director Elmer M. Ward, who had been instrumental in planning for this dinner, died on August 27. At the Annual Meeting following his death, he was posthumously given the Roy W. Crum Distinguished Service Award. These words were included in the award citation: "His efficient and untiring efforts have immeasurably advanced the cause of highway research and enhanced the stature of the Highway Research Board."

Following Mr. Ward's death, M. Earl Campbell was appointed Acting Assistant Director. He was also named Program Engineer for the National Cooperative Highway Research Program (which became a reality on June 19, 1962).

Ross Netherton assumed the duties of legal counsel in September for the newly established Special Committee on Highway Law, and Adrian Clary was employed on September 8, 1961, as Engineer of Maintenance, relieving Mr. Bollen of double duties (Materials and Construction plus Maintenance).

A conference was sponsored by the Special Committee on Urban Research on September 18-19 to outline priority problems of highway research in urban areas. One of the recommendations was the establishment of a department with a clearinghouse for urban research.

Topics discussed at the Annual Meeting included deterioration in concrete bridge decks, rolling bituminous surfaces with rollers and traffic, management improvement, fiscal control, aesthetics in highway design, corrosion-resistant steel, time-lapse analysis of construction methods, inhibitors of deicing salts for autos, treated wood in highway construction, effect of aging on asphalt, parking problems in planning, shoulders and lateral placement, lane use and turn controls, color aerial photography in material surveys, strength criteria for repeated loads, driving simulation, freeway operation, and rubber in asphalt.

R. R. Bartelsmeyer became Chairman of the Executive Committee in 1962. 1962

The AASHO Road Test field office closed in January 1962 and the remaining staff moved to Washington, D.C., to complete the final report and dissemination of data. William N. Carey, Jr., who had been named Technical Assistant at the January business meeting of the Executive Committee, assumed the office of Assistant Director during the summer.

Edward C. Squires joined the staff in May as a member of the administrative staff.

The Department of Economics, Finance and Administration held a conference on Planning in Highway Administration in March. A syllabus complemented the proceedings.

On June 19, strongly encouraged by A. E. Johnson of AASHO and E. H. Holmes of BPR, the National Academy of Sciences–National Research Council entered into a three-way agreement to be supported by participating member departments of the American Association of State Highway Officials. The National Cooperative Highway Research Program thus created (see Chapter 7) has as participating agencies the American Association of State

Highway Officials, the U. S. Bureau of Public Roads, and the National Academy of Sciences–National Research Council. Administration of NCHRP within the Highway Research Board was assigned to the first Program Engineer, M. Earl Campbell. Approximately $3 million annually was allotted to this research enterprise.

The Highway Laws Committee held its first workshop in July. The workshop was sponsored by HRB, BPR, and the University of Wisconsin.

John Guinnee was employed as Assistant Soils Engineer in July and Roy C. Edgerton as Assistant Engineer of Design in December.

Through funding by BPR, a clearinghouse was established by the Special Committee on Urban Research and the first issue of *Urban Transportation Research Digest* was published in Circular 489. Later issues were published in the *Highway Research News,* continuing through No. 10, published in March 1964.

Subjects of research interest in 1962 included continuously reinforced concrete pavement, insulation of bridge decks, protecting pavements from de-icing chemicals, traffic flow theories, rock slopes, visual needs for night driving, costs and benefits of highway improvement, values of time, stress distribution in soils, AASHO Road Test applications, serviceability index rating, graduate program in highway planning, programming highway accident reduction, repeated stresses in highway bridges, crash research, 30th peak-hour trend, highway bond financing, composite pavement design, skidding and road roughness, and new concepts in traffic assignment.

A conference on the AASHO Road Test was held on May 16-18 in St. Louis. Conclusions and interpretations were presented in a series of papers.

Herbert S. Fairbank, long a leader in highway research at BPR and involved as department chairman and in other roles with the Board for many years, died December 14.

1963

C. D. Curtiss served as Chairman of the Executive Committee in 1963.

On June 20, 1963, the Department of Urban Transportation Planning was organized from the Special Committee on Urban Research. Six project committees constituted the initial organization.

The Special Committee on Highway Laws sponsored its second workshop, held in June.

By 1963 three departments had nearly completed organizing into divisions: the Department of Economics, Finance and Administration; the Department of Design; and the Department of Soils, Geology and Foundations.

William Goodwin and Edward A. Mueller were employed as Project Engineers for NCHRP on January 1 and April 17 respectively. William

Gunderman was employed as Assistant Materials and Construction Engineer on June 3 and Robert C. Blensly took over the Economics, Finance and Administration Department in August.

With assistance from HRB in determining course content, West Virginia University inaugurated a graduate program in highway planning—the only program of its kind in the United States.

Old and new problems discussed in 1963 included adhesion and stripping of asphalt, chemical reactions in concrete, nondestructive testing of concrete, continuing evolution of photogrammetry, trip generation and traffic assignment, theory of traffic flow, and sonic energy in highway construction.

The year 1963 was the last under the administration of Mr. Burggraf, who became a consultant to the Board after the 43rd Annual Meeting in January 1964. A comparison of the situation at the beginning and end of his administration will show the dramatic growth made in 12 years:

Activity or Unit	*1951*	*1963*
Departments	6	7
Project committees	68	96
Special committees	1	8
Committee members (total men)	655	1,255
Committee assignments (aggregate)	950	2,651
Individual associates	1,118	1,528
Annual Meeting registration	850	2,443

During Mr. Burggraf's administration several outstanding accomplishments were noted: (a) completion of Maryland Road Test, (b) WASHO Road Test, (c) AASHO Road Test, and (d) initiation of the National Cooperative Highway Research Program.

The role of the Board in behavioral researches increased notably in this period.

The year 1964 brought about several significant changes and several important events. Wilbur Smith assumed office as Chairman of the Executive Committee. 1964

In January following the 43rd Annual Meeting, D. Grant Mickle succeeded Fred Burggraf as executive director. Mr. Mickle, who had served as Deputy Federal Highway Administrator since 1961, left that post to join HRB. His career, after graduation as a civil engineer from the University of Michigan and Harvard's Bureau of Street Traffic Research (the precursor of the Yale Bureau of Highway Traffic), included employment with the Massachusetts Department of Public Works, Michigan Department of State Highways, and the City of Detroit, and then with the Automotive Safety Foundation until 1961. For many years he had been highly active in HRB committee work, particularly in the Traffic and Operations Department and as Chairman of the Committee on Origin and Destination.

William N. Carey, Jr., who was serving as assistant director, was named deputy executive director and worked closely with Mr. Mickle, who served until September 30, 1966.

Mr. Mickle was adept as an organizer and aggressive in administration. He immediately began to formulate written statements of policies and procedures, pushing for clarification of administrative processes, establishing goals and seeking approaches to those goals through every feasible means. Delegating responsibilities and commensurate authority, he assumed the role of policy-maker and interpreter, leaving technological execution to subordinates. He involved himself with strategy but assured himself that the tactics were in responsible hands. A strong advocate of public relations, he augmented that office in the Board. Excerpts from Mr. Mickle's report to the Executive Committee a year after assuming office point up these facts:

The year through which we have just passed has been characterized by re-evaluations and adjustments of many facets of the Board's administration and activities.

Before I comment any further on this, though, I would like to express my sincerest thanks and deepest gratitude to the department and committee chairmen and to the committee members and HRB staff for their unstinting efforts in behalf of the Board and its mission. During my first year as Executive Director, I have been immeasurably impressed by their devotion and ability. These are the people who give the Board its dynamism and purpose, and without them any effort would be worthless. Their continued support and knowledge are the keystone upon which the Highway Research Board's future rests.

I would like to review some of the changes that have already been made and to preview the course that has been set for the Board to follow in the year to come and beyond.

Early last year the state highway departments, the Bureau of Public Roads, and other sponsors of the Board recognized that we needed increased financial support in order to enter into many new and vital areas of activity and service—a long-sought goal.

As a result of these increased funds we were, in the ensuing months, able to add personnel to increase the scope of our existing departments and to staff new departments to cover those areas that have been somewhat neglected in the past. At the same time, key Board personnel began a series of personal visits to the chief administrators of state and federal highway agencies to obtain, first-hand, advice and guidance on how our stepped-up activities and efforts might best serve their needs.

During the past year, also, the Board asked some 100 leaders of industries with an interest in highway transportation for their comments about how the Board might strengthen our already close ties. In return, we received some cogent and illuminating answers that will help us in the year ahead to increase our service to the Board's present industrial members and hopefully to attract other industries to join with us in a common effort to achieve the finest possible highway transportation system.

While nearly all departments were compiling statements of research needs, the ad hoc committee on Research in Behavioral Sciences stated in a report that, "Sociological, psychological, aesthetic, architectural, and similar research related to the provision of highways is important and long overdue."

Kenneth G. McWane, traffic engineer, retired because of ill health (about mid-year) and was succeeded by Edward A. Mueller, Projects Engineer of NCHRP.

Arthur B. Mobley joined the staff in July as RCS Coordinator. He later became Manager of the Highway Research Information Service.

The Department of Legal Studies was organized in January from the Special Committee on Highway Laws. In April it sponsored the third workshop on highway laws.

The first NCHRP Report, *Evaluation of Methods of Replacement of Deteriorated Concrete in Structures*, was published the middle of the year.

Later in 1964 Mr. Mickle requested the program engineer of NCHRP, M. Earl Campbell, to assist him as Engineer of Special Assignments. William A. Goodwin, who had served as a project engineer for physical research since January 1, 1963, succeeded to the position of Program Engineer, NCHRP, on October 15. Mr. Campbell was retained as NCHRP adviser.

The NCHRP staff underwent more changes when Carl Van Vechten was employed in July and when Howard Bissell and Krieger Henderson succeeded Mr. Mueller and Mr. Goodwin on August 23.

An event of special note was the start of the development period of the Highway Research Information Service (HRIS). This 3-year period began on March 16 as NCHRP Project 20-1 under the supervision of Dr. Paul E. Irick, Special Projects Manager, who had formerly been in charge of experiment design and data analysis at the AASHO Road Test.

In June 1964 James C. Allen joined the HRB staff. Mr. Allen had been a senior administrative officer of BPR for many years and was quite familiar with the complex relationships among NAS/HRB, the states, and BPR. He was also recognized as an outstanding manager of fiscal affairs. Mr. Mickle assigned Mr. Allen the responsibility for developing new budgetary and administrative procedures for the Board.

At its meeting in 1964 the Executive Committee requested an exploration to determine the desirability of creating a Special Committee on International Cooperative Activities. This exploration was made and a report given to the Executive Committee for its consideration in November.

At the 1945 Executive Committee meeting, the following was recorded: "It is the consensus of this meeting that public relations work of the sort described be undertaken, i.e., the dissemination of brief, common-language reports of research results, not to secure publicity for the Board but to convey important information to the public." It was not until November 1964, when David H. Buswell was employed, that a full-time public information officer could begin to fulfill this objective.

On December 5, 1964, the National Academy of Sciences established a sister Academy, the National Academy of Engineering. Both Academies were served by the National Research Council, the operating unit for research activity.

Problems discussed in 1964 included location by digital computer, interchange ramp color delineation, trends in auto ownership, rheological properties of soil-asphalt mixes, aerial surveillance techniques, highway ton-miles, nuclear applications in highway work, orthotropic deck sections, insulation to attenuate frost action, slip-form paving, space-use and traffic generation, highway benefits in developing countries, planning the metropolitan airport, measuring community values, fatigue strength of composite beams, snow and ice control by electric heating, planned communities, driving simulation, aluminum as culvert material, use of air space over highways, and basic research on concrete.

1965

Among the newsworthy activities recorded in 1965 was the authorization of the Special Committee on International Cooperative Activities in January by the Executive Committee, in response to the desire of many highway engineers in countries abroad to have a closer rapport with the Board. Wilbur Smith, a man with worldwide consulting interests, was named Chairman. Donald S. Berry succeeded Mr. Smith as Chairman of the Executive Committee.

The committee's objectives were to continue and expand current HRB international cooperative activities; to translate and abstract from major foreign-language publications selected research and development reports for use by the Board's automated Highway Research Information Service; to provide sessions at the Board's Annual Meetings for presentation of papers dealing with highway research activities abroad; to encourage HRB staff and committee participation in international research conferences; to expand contacts and information exchanges with the Board's 37 counterpart organizations in other countries; and to incorporate into the HRIS material gathered through terms of reciprocal agreements made with other appropriate organizations.

James A. Scott was employed on January 1, 1965, as Urban Transportation Planning Specialist. He succeeded M. Earl Campbell in the Department of Urban Transportation Planning.

In February a "Colloquy on Motor Vehicle and Traffic Law" was held by the Department of Legal Studies. The workshop on highway laws had by now been established as an annual event. The Department of Legal Studies also began a new series entitled *Highway Law Comment.*

In June E. W. Harris left the position of Assistant to the Executive Director to join the Executive Offices of the National Academy of Sciences as Personnel Officer for the Academy.

The new *Highway Capacity Manual,* which had been in preparation for 7 years in cooperation with BPR through their Committee on Highway Capacity, was published. This monumental manual (400+ pages) was dedicated to the late O. K. Normann, who had chaired the committee until his death on May 4, 1964.

The first issue of *Highway Research In Progress,* an experimental output from the Board's HRIS, was published in September. The same month the BPR issued *National Program of Research and Development for Highway Transportation* (BPR Circular). This was to serve as a guide for BPR and for cooperative endeavors.

In October Congress passed the Highway Beautification Act—an act that stimulated action in several of the Board's departments. One result was the publication of Special Report 88, *The Art and Science of Roadside Development,* by the Department of Design.

In November the Department of Urban Transportation Planning arranged a "Workshop on Land-Use Models: Planning, Application and Research Needs."

Also in November L. F. Spaine and William Williams joined the NCHRP staff.

The Board, cognizant of the need for a comprehensive, continuing program of research directed toward national and state goals, had begun a framework study that was to be a cooperative endeavor by all departments (and committees) working with the Board's staff.

A new Special Committee on Long-Range Planning, authorized by the Executive Committee, was appointed on December 17. The committee's objectives were to examine the broad concepts and needs of total transportation relating to new technology and methodology and identitfy those for which Highway Research Board activity would be appropriate; to review the functions, structure, and finances of the Highway Research Board with the object of defining desirable changes; and to evaluate existing research activities and programs from the viewpoint of future needs for transportation research.

Topics of concern during the year included current techniques to shape the urban form, Interstate Highway shoulder use, control of traffic in tunnels, studded tires, asphalt composition related to behavior, travel trends in metropolitan areas, measurement of highway benefits, soil classification systems, skid-resistance guidelines, speed-density analyses, piggyback and containerization in transport, highway functions and amenities, upgrading of line-haul trucking costs, rigid, flexible, and composite pavement design, temperature-flow phenomena in asphalts, the Benkelman beam, and inventorying research in foreign countries.

1966 In January 1966 Fred Burggraf, former Director of the Board, died. At the time of his death he was working on the history of the Highway Research Board, which he knew so well and of which he had been a dedicated and loyal employee for so many years. Of him a friend said, "He worked two days every day with prodigious energy and enthusiasm. A perfectionist in his profession, he read everything the Board prepared for publication to make doubly sure that it was ready. Withal he was a warm-hearted, out-giving friend, devoted Lutheran and family man." The Fred Burggraf Award was created in memory of the late HRB director. It is given annually to recognize Annual Meeting papers of outstanding merit by researchers 30 years of age or younger.

J. Burch McMorran was elected Chairman of the Executive Committee.

Stephen Montgomery, who in 1969 became Senior Editor of HRB publications, joined the staff in January as Assistant Editor.

In February 1966 the Division of Engineering and Industrial Research of the National Research Council resumed its former name, Division of Engineering.

On April 1 Special Project 20-2 was awarded to Tallamy and Smith. This project, "Research Needs in Highway Transportation," resulted from the recognition by the AASHO Evaluation Committee that the existing method of programming NCHRP projects was in need of stated objectives. There was need to establish goals and program projects within the framework of these goals in a continuing, comprehensive program for NCHRP.

The Board, which had begun to design a framework study for a nationwide program of research, furnished the contractor the information assembled, including some 900 project statements.

HRB departments and committees were at work abstracting monumental research papers and reports for storing in the HRIS computer memory.

Franklin N. Wray, Engineer of Design, retired on June 30 and was succeeded by his assistant, Roy C. Edgerton.

In September 1966 Louis Jordan retired as Executive Secretary of the Division of Engineering of the National Research Council. He had served in that position since September 1948, and, in that capacity, was a member of the Executive Committee of the Board during the entire period. He was an enthusiastic supporter of the Board in the senior councils of the Academy and a wise and sympathetic adviser to the staff.

On September 30 Executive Director Mickle left HRB and returned to the Automotive Safety Foundation to become Vice President. During his nearly 3 years with the Board there had been a healthy expansion in service activities, a tightening and sharpening of the organization in keeping with its objectives and functions, and a delegation of responsibilities. A "charter of accountability" had been effected.

William N. Carey, Jr., succeeded to the position of Executive Director. A civil engineering graduate of the University of Minnesota in 1937, Mr. Carey worked in cement research in Minnesota and on airport construction during the war before joining HRB. A veteran with the Board, he had served in various capacities since employment on January 21, 1946. Executive Assistant to the Director until 1952, he subsequently became Project Engineer for the WASHO Road Test and Chief Engineer for Research on the AASHO Road Test. He became Assistant Director in 1962, and then Deputy Executive Director in January 1964.

On November 1 Mr. Carey announced the following appointments: Roy C. Edgerton, Assistant Director for Technical Activities; Paul E. Irick, Assistant Director for Special Projects; James C. Allen, Assistant Director for Administration; William A. Goodwin, Program Director for NCHRP; and Lawrence F. Spaine, transferred from NCHRP to become Engineer of Design in the Technical Activities Division.

In October the Department of Transportation was created in the federal government, combining in one department the government's concerns for highway, rail, and air transportation. Several states underwent a similar restructuring. These developments gave increasing emphasis to the need for research into modal interactions.

In December Harry A. Smith joined the NCHRP staff as a Projects Engineer.

Topics discussed in 1966 included thin surface treatments, hydroplaning of autos, accident rates related to geometric design, pneumatic roller compaction of asphaltic concrete, tungsten-carbide tipped snowplow blades, breakaway signs, traffic assignment with capacity restraints, skid testing with an automobile, highway pollution problems, compaction of earthwork and bases, highway illumination and delineation, gap acceptance in freeway merging, truck fuel consumption related to geometrics, subsurface drainage, application of statistical concepts in accident data, Interstate System accident research, road capacity of city centers, driver communication, realistic minimum standards, or levels of service, for use in functional classification, and protecting steep slopes against erosion.

1967 At the January 1967 Annual Meeting there were 3,311 registrants. Of these, 200 were from 17 foreign countries, indicating the worldwide interest in Board activities.

Edward G. Wetzel succeeded J. Burch McMorran as Chairman of the Executive Committee.

The Board undertook a special study of bus transportation for the Department of Housing and Urban Development. Byron Bledsoe rejoined the staff in March to serve for a year as principal investigator.

In May the Traffic and Operations Department held a 3-day conference on Improved Utilization of Streets Through Traffic Engineering.

On June 30 A. W. Johnson, Soils Engineer since January 1, 1946, retired because of ill health. He was succeeded by John W. Guinnee, his assistant for 4 years. On August 27 Mr. Johnson died.

On July 1, HRIS completed its 3-year developmental stage and became a part of the functional operations of the Board. The second edition of *Highway Research In Progress* was published in May; it contained 5,000 abstracts of research, both domestic and foreign. HRIS also published a *Current Awareness* report, providing up-to-date information on specific subjects, and provided system searching for the provision of all information stored on specific subjects.

Toward the latter part of 1967 and as authorized by the Executive Committee, the Board created a Special Committee on the Conduct of Research, a committee on research into research. It was dedicated to improvement of management, organization, performance, and implementation of highway research.

William A. Goodwin resigned on August 31 from directorship of NCHRP and returned to the University of Tennessee. He was succeeded by Krieger Henderson, who had served as Project Engineer since August 1, 1964.

M. Earl Campbell retired on September 30 and returned to West Virginia to serve as Assistant to State Highway Engineer (Planning, Administration, and Maintenance), West Virginia Department of Highways. Kenneth E. Cook took over the Department of Economics, Finance and Administration in September and James R. Novak joined NCHRP in October as a Projects Engineer.

In November the executive director and the chairman of the Special Committee on International Cooperative Activities attended the Tokyo meeting of the Permanent International Association of Road Congresses as representatives of the Board, which had renewed its affiliation with PIARC in 1967.

The Long-Range Planning Committee brought its report to the Executive Committee in January 1967. The Executive Committee approved the following recommendations:

Purpose—The purpose of the Board is to advance knowledge concerning the nature and performance of transportation systems, through the stimulation of research and dissemination of information derived therefrom.

Scope—The Board will give attention to all factors pertinent to the understanding, devising, and functioning of highway and urban transportation systems and their interrelationships with other aspects of total transportation.

It will concern itself with the planning, design, construction, operation, maintenance, and saftey of facilities and their components; the economics, financing and administration of the systems; and their interactions with the physical, economic, legal, and social environment they are designed to serve.

Functions—The purpose of the Board will be pursued through activities designed to (a) identify research needs and resources; (b) stimulate research efforts; (c) provide for the exchange of concepts and techniques relating to research; (d) foster the reporting, evaluation, synthesis, and interpretation of research findings, and provide for their publication where appropriate; (e) promote coordination of national and international research efforts; and (f) encourage the dissemination and utilization of knowledge derived from research.

The statement of purpose and scope was approved by the Division of Engineering in August 1968. It was submitted to the Governing Board of the National Research Council as information in October 1968 and approved by the Governing Board April 27, 1969.

The new purpose and scope did not in fact change the activities of the Board, but rather reflected the broadening of direction and activity that had been taking place within the Board's committees. The major functions of the Board would be accomplished through the voluntary efforts of those associated with activities of the Board, enhanced to the extent feasible by efforts of a staff and to the extent practical under support that may be made available.

Special research projects would be conducted when such projects had been approved by the Executive Committee and accepted by the National Academy of Sciences as appropriate to its function.

Advisory, correlation, publication, or other services would be rendered when such services had been approved by the Executive Committee and financed in accordance with the policies of the Board and the Academy. Funds provided by sponsoring agencies would be used to support those activities stipulated by the agreement with the sponsors.

Topics of concern in 1967 included perspective plotting of highways, Interstate Highway maintenance requirements, vision and driving, the rapid-travel road profilometer, painting of highway steel, pavement thickness determination by ultrasonic pulses, modal choice, planning and programming of maintenance, demand for air travel, standards for width of low-volume roads, skid resistance and surface treatments, ground transportation at major airports, the public's attitude toward transport mode, team concept in urban highway design, anti-skid materials on ice- and snow-covered pavement, multiple use of highway rights-of-way, effects of admixtures in concrete, optimization of traffic-flow splits, trip generation, pavement service lives, travel estimation models, effects of loadings on bridge life, value of commuters' time, and colored pavement for traffic guidance.

Lowell K. Bridwell was Federal Highway Administrator in 1967 and 1968. He first became interested in the activities of the Board at the time of the AASHO Road Test, and was an enthusiastic and effective supporter of the Board during his tenure as Federal Highway Administrator. He continues to maintain his interest as a member of the Division A Council.

1968 David H. Stevens became the new Chairman of the Executive Committee in 1968, and in May, pursuant to the Executive Committee's approval of the new purpose and scope proposed for the Board, a conference on reorganization and reorientation was held at Airlie House in Warrenton, Virginia. One of the missions of the conference was to develop a more detailed organizational structure to accommodate those existing HRB activities that should be continued and those additional activities implied by the new scope and purpose that should be undertaken now or in the future.

Those invited to participate in the conference were selected from among the most active participants in the work of the Board. They were men who had long tenure in committee activity, who fully understood the Board's operational mechanisms, and who recognized the directions the Board must take to be responsive to the needs of the future. Other delegates were men with a broad approach to transportation who were interested in counseling with highway specialists.

At the conference a new organizational framework conceived by the Board's Long-Range Planning Committee was endorsed by the delegates. This framework consisted of four main Divisions within the Board:

Division A—Regular Technical Activities
Division B—Special Technical Activities
Division C—Administrative Activities
Division D—National Cooperative Highway Research Program Administration

This structure was basically the same as the structure then being followed; however, Division A would include three Groups instead of the eight Departments existing under the old structure. The three Groups are as follows:

Group 1—Transportation Systems Planning and Administration
Group 2—Design and Construction of Transportation Facilities
Group 3—Operation and Maintenance of Transportation Facilities

The reorganization was to be accomplished in stages, with a target date of January 1970 for completion.

The Traffic and Operations Department held a workshop in January 1968 on "Human Factors in Highway Transportation" and a symposium in May on "Visibility and the Driving Task."

The Department of Materials and Construction instituted an *Information Series* on the latest research developments.

Topics discussed at the 47th Annual Meeting included tire hydroplaning and tire traction, effect of fares on transit riding, demand for air travel, night driving and vision, design criteria for breakaway sign supports, impact behavior of luminaire supports, grade separations on rural freeways, influence of water content on concrete consistency, value of commuters' travel time, and chemical expansion of fresh concrete.

The NCHRP *Summary of Progress* through June 30, 1968, noted 121 contracts let, 54 reports published, $13.5 million obligated, and 17 additional projects being processed for contract. In November, NCHRP Report 55, *Research Needs in Highway Transportation,* was published. A quotation from this report seems fitting:

> To utilize best the efforts made in the field of highway research, and to make those efforts a coordinated part of an enlightened transportation research program, it is desirable that the highway transportation research goals be promoted as those toward which all research in this field will be focused. With these goals as objectives, highway research in general will have cohesiveness and focus to ensure its compatibility and direction; automatically there can be a coordinated national research effort in the field of highway transportation.

At the suggestion of the Institute of Transportation and Traffic Engineering at the University of California and the Civil Engineering Department at the Polytechnic Institute of Brooklyn, and with the support of the Automotive Safety Foundation, the Board convened a conference on "Joint Development and Multiple Use of Transportation Rights-of-Way" in November. The conference was attended by an interdisciplinary group of professional and community leaders. Several case studies were presented, including those of such urban areas as Baltimore, Chicago, Los Angeles, New York, and Minneapolis-St. Paul. The proceedings of this conference were published in HRB Special Report 104.

During 1968 Louis MacGregor and W. C. Graeub joined the NCHRP staff—MacGregor as Administrative Engineer, Graeub as a Projects Engineer. Thomas Copas took over the synthesis project sponsored by NCHRP, and C. Kim joined the Special Projects branch as a Senior Computer Programmer.

1969

Oscar T. Marzke was Chairman of the Executive Committee in 1969.

On February 20, 1969, Pyke Johnson died at the age of 80. He had been involved with the Board from its very beginning, having attended the organizational conference of November 11, 1920, in New York City. Mr. Johnson maintained an active interest in the Board up to the time of his death. His contribution as a counselor to the staff and Executive Committee went far beyond the work of the committees on which he served.

One of the milestones of 1969 was the initiation of the Transportation Research Information Service (TRIS I) by the Special Projects staff, under the direction of Dr. Paul Irick. The proposal for TRIS I had been accepted by the Department of Transportation Office of Industrial Cooperation in December 1968. This proposal provided for funding by that office for one year; then in March TRIS I received additional funds from the DOT Office of Research and Technology. Later in the year DOT extended the project time and increased its funding. Also, the Maritime Administration initiated still another information service—the Maritime Research Information Service (MRIS)—to be administered by Special Projects in cooperation with the Maritime Information Committee of the National Research Council.

Attendance at the 48th Annual Meeting reached a high of 3,284 registrants. Subjects under discussion included ground transportation facilities at airports; attention value of highway signs; nuclear depth gage study of subgrade moisture; computer graphics and visual highway design; analytical aerial triangulation; geology, cities, and surface movement; route guidance; tire disablements and accidents; the effect of modern highways on urban manufacturing growth; and relocation problems in a major city.

The Highway Research Board conducted a conference in May on

"Transportation and Community Values" sponsored by the Bureau of Public Roads and the Department of Housing and Urban Development, with the cooperation of the Urban Mass Transportation Administration. Here again the NRC Division of Behavioral Sciences worked closely with the Board in program formation and the identification of participants. This meeting emphasized the interrelationship between transportation planning and the community affected by such planning. The proceedings of this meeting were published in HRB Special Report 105.

In April the Executive Committee approved the revision of the Bylaws, which now expanded that committee to include 28 members. These Bylaws also provided for the new structuring of the Board with its four divisions, to be put into effect in 1970.

On May 7-9 a conference was held in Warrenton, Virginia, to study transportation of hazardous materials. This conference was the cooperative effort of the NRC Division of Chemistry's Committee on Hazardous Materials and the Board, under contract to the Office of Hazardous Materials of the Department of Transportation.

In 1969 the Board sponsored the Second Western Summer Meeting in Salt Lake City, Utah. This meeting, cosponsored by the Utah State Department of Highways, presented timely and useful information on many facets of the overall theme of highway safety. Such topics as safety and highway design, the modular crash cushion, driver education and licensing, driver communication through roadway delineation, and roadside structures and protective systems were discussed. The success of this meeting, attended by over 200 registrants, was the determinant in scheduling the Third Western Summer Meeting in Sacramento, California, for the summer of 1970.

In 1969 Ross Netherton left the Board and was succeeded by John Vance in July. K. B. Johns joined the Technical Activities staff in February. Elizabeth Cole, a librarian, and Rosemary Mapes, an editor, joined the staff in 1969. Senior Systems Specialist H. S. Schofer joined the Special Projects Division to help in the development of the Board's many information systems. David Oliver was employed in the legal research area in August.

Frederick Seitz, President of the National Academy of Sciences since 1962, left that post in 1969 to become President of The Rockefeller University. During his term as President of the Academy, Dr. Seitz had actively supported the Board in many ways, most notably through his inspiring messages at the Board's Annual Meetings. Dr. Seitz was succeeded in July 1969 as President of the Academy by Philip Handler. Dr. Handler has continued the tradition of strong interest in the Board's activities shown by the Academy's senior officers.

1970 Years of planning and study bore fruit on February 1, 1970, when the reorganization of the Highway Research Board's committee structure went into effect. This had been the subject of discussion for several years, and became an accomplished fact only 9 months after the new purpose and scope were approved by the Governing Board of the National Research Council.

As a result of the reorganization, the Board was now charged with the responsibility of advancing knowledge concerning the nature and performance of all forms of transportation, especially as they relate to highways.

The international appeal of the Board's Annual Meeting was demonstrated in January when 171 of the 3,000 registrants came from overseas, representing 15 different countries. Sixty technical and conference sessions ran the gamut of research from soils to safety.

A familiar face in a new job resulted from the meeting. D. Grant Mickle, past Executive Director of the Board, was installed as the Chairman of the Board's Executive Committee, only a few days after he was appointed to the full-time position of President of the newly formed Highway Users Federation for Safety and Mobility.

In April 1970 David H. Buswell, Special Assistant to the Director, resigned the post that he had held for 6 years, and Hugh M. Gillespie, previously Director of Public Relations for the National Asphalt Pavement Association, replaced him in the capacity of Public Information Officer.

National concern in the area of noise pollution was reflected in the formation of the Transportation Noise Research Information Service (TNRIS), which was sponsored by the Office of Noise Abatement, Office of the Secretary, Department of Transportation. Headed by a four-man staff from HRB, the new service was initiated to collect and disseminate information on noise and vibration in various modes of transportation. Over 2,000 references to on-going research projects and published articles were due to be processed in the first year as a basis for further work.

A. A. Jones, Herbert Pennock, James Wright, and Cloyd V. Taylor joined the staff of HRB during the year.

While the Board prepared to celebrate its 50th Anniversary, a different kind of accomplishment was observed by the National Cooperative Highway Research Program. During 1970 the number of reports published by NCHRP since its formation 8 years earlier passed the 100 mark.

Following in the footsteps of the Department of Legal Studies, the new Legal Resources Group conducted a double-barreled event that had been initiated with great success the previous year—the "Institute on Motor Vehicle and Highway Law" at the University of Colorado, and the "Workshop on Highway Law" at the University of North Carolina.

Group 2, Design and Construction of Transportation Facilities, planned and conducted a workshop on the "Structural Design of Asphalt Concrete Pavement Systems," sponsored by the Federal Highway Administration and held in Texas during December.

Another workshop, developed by Group 1, on "Transportation of Goods in Urban Areas," was particularly significant because it marked the first HRB effort with joint financial sponsorship of the U. S. Department of Transportation and the Canadian Department of Transportation. The workshop was held in December.

Much interest was aroused by a special workshop sponsored by the Maintenance Department during April on the subject of "Snow Removal and Ice Control Research." This was held at the U. S. Army Cold Regions Research and Engineering Laboratory in Hanover, New Hampshire.

Among subjects of more than usual interest under discussion and consideration by HRB during 1970 were public transportation to airports; natural versus synthetic aggregates; aggregates and skidding; alcohol, alertness, and drug abuse; and statistical procedures in research.

At the 49th Annual Meeting in January 1970, Francis C. Turner was awarded the Roy W. Crum Distinguished Service Award in recognition of his outstanding contribution to highway progress through research. Mr. Turner was appointed Federal Highway Administrator in February 1969, following a distinguished career with the Bureau of Public Roads that began as a junior engineer in 1929 and culminated with his appointment as Director of the Bureau in February 1967. He has been an active supporter of Board activities, primarily through his encouragement of BPR staff members who were participating in Board activities.

CHAIRMEN OF THE EXECUTIVE COMMITTEE HIGHWAY RESEARCH BOARD 1920-1970

Anson Marston
1920-22

A.N. Johnson
1923-26

Thomas R. Agg
1927

Frank H. Eno
1928-29

H.S. Mattimore
1930-31

George E. Hamlin
1932-33

Albert T. Goldbeck
1934-35

H.C. Dickinson
1936-37

Burton W. Marsh
1938-39

W.W. Mack
1940-41

Fred C. Lang
1942-43

Stanton Walker
1944-45

Roger L. Morrison
1946-47

F.V. Reagel
1948-49

Ralph A. Moyer
1950-51

R.H. Baldock
1952-53

Walter H. Root
1954

G. Donald Kennedy
1954-55

Kenneth B. Woods
1956

Rex M. Whitton
1957

Charles S. Scholer
1958

Harmer E. Davis
1959

Pyke Johnson
1960

W.A. Bugge
1961

R.R. Bartelsmeyer
1962

C.D. Curtiss
1963

Wilbur S. Smith
1964

Donald S. Berry
1965

J. Burch McMorran
1966

Edward G. Wetzel
1967

David H. Stevens
1968

Oscar T. Marzke
1969

D. Grant Mickle
1970

4

Structural Evolution of the Highway Research Board

Beginning as the Highway Research Committee in the Division of Engineering of the National Research Council on October 31, 1919, the structure of the organization has passed through several stages in its evolution. As growth continues and expansion of concern and service continues, the structure must also continue to change.

For simplification this evolution will be treated here in two parts. *Evolution of Administrative Structure* is concerned with the Executive Committee and the administrative staff, embracing the changes from the embryo Highway Research Committee to formation of the Advisory Board on Highway Research (1920), to the Highway Research Board (1925) with its Research Correlation Service (1945), National Cooperative Highway Research Program (1962), Special Tasks (1962), Highway Research Information Service (1964), and International Cooperative Activities (1965).

Evolution of Technological Structure is concerned with the technical committee evolution, beginning with the three subcommittees of the Highway Research Committee established in 1919 and their increase to seven by 1930, when five sections with 33 committees were established, through to the creation of seven departments with 35 committees in 1935, to the eight departments with 110 committees and an additional seven special committees in 1968, to the consolidation in January 1970 into one Division of Technical Activities consisting of three groups.

The separation of this chapter into two parts is oversimplified because the Board operates as a unit. Administration pervades all activities, and technical activities are the essence of the organization. Yet for an analysis of the structure, this two-part division provides a straightforward treatment.

Evolution of Administrative Structure

The Highway Research Committee, authorized in the Division of Engineering on October 31, 1919, consisted of four persons, Dean Anson Marston (Chairman), Professor A. N. Talbot, H. H. Porter, and G. S. Webster. Three technical subcommittees were administered by the parent committee. This arrangement continued for a year.

Executive Committee

On November 11, 1920, the Highway Research Committee, with its three technical subcommittees, was reorganized into the Advisory Board on Highway Research. An Executive Committee was established with five members. Dean Marston continued as Chairman, and Alfred Flinn, Vice Chairman of the Division of Engineering, was made Vice Chairman of the Executive Committee. Other members were Thomas H. MacDonald, George S. Webster, and Charles F. Kettering. Provision was made for eight research committees, including the three already in existence. Of the eight named, only five were then established, and many years passed before more were established. An amendment on August 4, 1923, increased the size of the Executive Committee from five to seven.

Although the Executive Committee on December 8, 1924, authorized the name of the Board to be changed on January 1, 1925, from the Advisory Board on Highway Research to the Highway Research Board, there was no other change in the Bylaws or structure at that time. An amendment on December 14, 1928, increased the size of the Committee to 11 and provided for staggered 3-year terms.

On August 1, 1945, the Executive Secretary of AASHO was designated an ex officio member of HRB. Previously, ex officio members were designated from the Bureau of Public Roads and the NRC Division of Engineering.

On December 21, 1956, the size of the Committee was increased to 19, and on September 27, 1957, the size was increased to 21. The two immediate past chairmen were to serve as ex officio members.

The Bylaws approved July 27, 1959, and printed in Volume 38 of the *Proceedings* provided for 21 members on the Executive Committee and spelled out functions of the five standing subcommittees of the Executive Committee that had been instituted during the chairmanship of Professor K. B. Woods. These subcommittees were made up of Executive Committee members and were designed to make Executive Committee activities more systematic and efficient. The five subcommittees were:

1. Research Needs and Project Initiation (8 to 12 members): The chairman would be the newly elected second vice chairman of the Executive Committee.
2. Publications Policy (8 to 12 members): The chairman would be the newly elected first vice chairman of the Executive Committee.
3. Research Interpretation and Dissemination (8 to 12 members): The chairman would be the immediate past chairman of the Executive Committee.
4. Ways and Means (8 to 12 members): The chairman would be the second immediate past chairman of the Executive Committee.
5. Steering (5 members): The members would consist of the chairmen of the other standing committees together with the newly elected chairman of the Executive Committee as chairman.

On February 1, 1961, the size of the Executive Committee was increased to 25 members, and it was stipulated that a member could not serve more than two 3-year terms consecutively except for ex officio members, who would serve as provided in the Bylaws.

With the Bylaws approved in 1969 came a slight change in the structure of the Executive Committee. Now a committee that could have as many as 28 members would provide the many and varied disciplines of concern in the highway transportation field. Also at this time the notion of five formal subcommittees was dropped in favor of a procedure in which all business is conducted by the full Executive Committee. As before, the officers of the Board were the chairman, vice chairman, the director, and his deputy (or designated assistant). Appendix G contains the current Bylaws approved April 27, 1969.

Research Correlation Service

Inasmuch as many pages are devoted to the Research Correlation Service elsewhere (Chapters 3 and 5), only the functions of the Service, the means of implementing the functions, and the modifications in structure of the Board to accommodate this new function are discussed here.

On June 13, 1945, "to give effect" to the recommendations and plans of the American Association of State Highway Officials, Herman MacDonald addressed a letter to each of the state highway departments advising that "a formal request for your subscription, payable July 1, 1945, will be sent you shortly by the Director of the Highway Research Board." Thus began the implementation of the concept. By the end of 1945 more than two-thirds of the state highway departments had subscribed to the correlation service. Since the late 1940's, with very few exceptions, all states and the District of Columbia and Commonwealth of Puerto Rico have subscribed.

The intent of the Research Correlation Service is to provide linkages:

1. Between the subscriber and all other organizations conducting highway research, including state highway departments, federal bureaus, municipal research units, colleges, industries, and associations. This is accomplished through HRIS and through personal visits by RCS specialists in the several professional disciplines in the highway transportation field. Each specialist visits all state highway departments, all state universities, and highway-related industries.
2. Between the subscriber and the comprehensive highway research program under the aegis of the Board, including the work of project committees, special tasks conducted by the Board, and research by the RCS staff and by NCHRP.

In addition to these special linkages the subscriber is provided other special service opportunities, including assistance in formulating research projects, preparation of special bibliographies through library research, compilation of operating practices, provision of lectures or seminars, and other types of appropriate services. In addition, RCS provides HRB publications. These services are accomplished largely through the efforts of the RCS specialists, with the help of the HRIS staff and other HRB staff.

When an RCS specialist visits an organization he has eight objectives in mind, the bases of correlation, collection, and dissemination:

1. Problems—discussion of current research needs or proposal statements for referral to departments or NCHRP.
2. Program—review of the organization's research program.
3. Projects—description of current research projects.
4. Progress—project-by-project progress report.
5. Papers—notation of possible papers for a forum.
6. People—identification of new men of competence for HRB committees and panels.

7. Practice, procedures, policies, etc.—results from application of research findings.

8. Payoff—account of benefits and costs of research in expenditures and returns.

When the specialist is in the office he is busy at several tasks, including (a) scanning trade and technical periodicals and research reports for important findings to refer to HRIS; (b) answering requests for information; (c) assisting committees in research and reports; (d) assisting in preparation and conduct of seminars and interim and annual meetings; and (e) preparing papers and reports for publication.

The Board was restructured in 1946 to provide for this Service. Five specialists with secretarial help were employed. Now the RCS staff has grown to 10 specialists with complementary secretarial staff.

In recent years RCS activities have been conducted in the Technical Activities Division of the Board and the name Research Correlation Service is now seldom used. The concept is the same, the activities are the same, and most of the people involved are the same. The only reason for change is that some feel that Technical Activities is more descriptive of what is involved and actually includes Research Correlation Service as well as other things.

The Reference Library

In 1946, after the establishment of the Research Correlation Service, a librarian was added to the staff to develop a research library that would assist the staff in the work of this program. Except for two short leaves, Mrs. Dorothy Bright has served as Librarian since December 1946.

The library was assigned space in the NAS building with other Highway Research Board offices, and the nucleus of a collection was assembled, consisting of the Board's own publications plus other documents and reports of general use. The collection was organized and card catalogs were developed to provide access to the information. Relations were established with outside information sources, such as the Bureau of Public Roads library and other governmental and private libraries in the area.

In 1951 a major responsibility was added to the work of the library. Previously the selection of material for *Highway Research Abstracts* had been handled by the office of the assistant director. At this time, the task was transferred to the Librarian.

The HRB staff, which had been split in 1956 because of space shortages at the Academy Building, was housed together in 1958 when new quarters were provided at 1707 H Street N.W. At this location more space became available for the library. The Librarian maintained the library, with occasional assistance from the secretarial staff, until 1961, when a full-time clerk-typist was added to the library staff.

In 1964, with the establishment of the Highway Resarch Information Service as another technical service of the Board, the library assumed new responsibilities in supplying information and material for this operation.

In 1965 the work load of the library had increased to such an extent that an additional staff member became necessary. An assistant librarian was added to the staff in November of that year.

With the transferral of HRB offices to the Joseph Henry Building at 2100 Pennsylvania Avenue N.W. in the fall of 1967, the library was located in quarters specifically planned for library operations, with space for possible future expansion. More book stacks and new equipment made it possible to provide more efficient service to the staff.

The library is a unit of the Technical Activities staff, under the direct supervision of the Assistant Director for Technical Activities. Its purpose is to provide all library services as required by the Board. These include (a) provision of professional library services in the planning and development of HRB activities that include bibliographic or other informational procedures related to library functions; (b) maintenance and operation of the Board's own specialized library facilities; and (c) awareness of the subject content of other libraries and other institutions, both in the Washington area and elsewhere, and maintenance of relations with these sources so that their materials may be accessible to the Board as needed.

The library consists of approximately 9,000 items, including:

1. Copies of all HRB publications and other informal reports. (Appendix H lists the publications series of the Highway Research Board.)
2. A collection of research reports and documents published by state and federal agencies, universities, industries, and associations working within the broad fields of highway transportation and technology; proceedings of conferences; and specifications and manuals of state highway departments.
3. A collection of approximately 225 periodicals.
4. Reports and periodicals received on an exchange basis from agencies in foreign countries.
5. HRIS Document Records for use by the staff and visitors.

Three card catalogs are maintained: one for HRB publications, one for the general library collection, and a subject catalog to material indexed in *Highway Research Abstracts*.

National Cooperative Highway Research Program

On June 19, 1962, a three-way agreement was signed by the National Academy of Sciences, the Bureau of Public Roads, and the American Association of State Highway Officials—an agreement that inaugurated a research program involving over $3 million a year in problem-solving for the state highway departments. The National Cooperative Highway Research Program is sponsored by participating member departments of AASHO, in cooperation with BPR, and is administered by the Board. Annually, AASHO proposes specific research projects that have been identified by highway departments and AASHO committees for inclusion in the National Cooperative Highway Research Program. The Program is discussed more fully in Chapter 7.

Special Projects

In 1962 the Board established the Special Projects unit with Paul E. Irick as Manager. This unit was to undertake in-house research as referred to the Board by AASHO and other appropriate sponsors. Special Projects activities are discussed in Chapter 7.

The Highway Research Information Service

In 1920 the purposes of the Board were stated as follows:

> To prepare a comprehensive national program for highway research; to assist existing organizations to coordinate their activities therein; and to collect and distribute information of completed and current research.

Through the years the latter part of this statement has been reiterated. In 1923 Director Hatt stated: "A highway research information service should be given careful consideration." In 1928 Director Crum requested $8,000 for a highway research information clearinghouse. In partial fulfillment of that request, *Highway Research Abstracts* was inaugurated in 1931. Periodically through the years, censuses of research in progress were made and reported through various publications.

Hence the inauguration of the Highway Research Information Service (HRIS) in July 1967 was not a new concept. Its automated mechanics were the new concept, and by means of the computer, a greatly expanded clearinghouse function was provided.

The 3-year design and development period, sponsored by the state highway departments and the Bureau of Public Roads at a cost of $800,000, was under the direction of Paul E. Irick, Assistant Director for Special Projects. Now that the system is operational it is in the Technical Activities Division. Arthur B. Mobley is System Manager.

Items selected for HRIS processing are in two general categories: (a) research in progress reports—one-page statements of project facts and objectives, about 10,000 in HRIS files (December 1970); and (b) abstracts of published articles or reports—about 28,000 contained in HRIS files (December 1970).

These items, classified into 34 subject areas, are increasing at the rate of 10,000 each year. Items are indexed by author, subject, and authorized term. From computer storage, HRIS provides abstracts of published works and summaries of on-going research projects in response to specific inquiries. HRIS also compiles information for two publications: *Highway Research In Progress* (annually) and *HRIS Abstracts* (quarterly). *Current Awareness* reports are issued monthly. Those requesting the current awareness reports specify their interests, and these are matched against recent HRIS acquisitions.

Research in progress reports are acquired either through the field visitation program of the HRB Research Correlation Service, through the Bureau of Public Roads, through service and exchange agreements with other information services, or as new reports furnished by research agencies during annual updating operations.

Approximately 200 U.S. research agencies are reached by field visits of the Technical Activities staff to state highway departments, universities, and other highway research groups. The HRB committees are another source for publications abstracts.

The Bureau of Public Roads provides HRIS with abstracts of reports from all BPR-sponsored research projects. The corresponding full-length documents are stored in the Clearinghouse for Federal Scientific and Technical Information. These reports are published annually in *Highway Research and Developmental Studies* by BPR.

Among the exchange agreements maintained by HRIS is one with the American Society of Civil Engineers. HRIS stores selected abstracts provided through ASCE services and provides ASCE with abstracts of HRB publications. Another such abstract exchange is that between HRIS and the National Safety Council's Safety Research Information Service (SRIS). HRIS also conducts an exchange with the Science Information Exchange of the Smithsonian Institution, an information service for research in progress in all areas of science and technology. A preliminary exchange is being developed between HRIS and the Transportation Association of America.

The major source of non-U.S. projects is the world survey of current research and development by the International Road Federation (IRF), in cooperation with the Organization for Economic Cooperation and Development (OECD). OECD is developing a storage and retrieval system for worldwide coverage of published highway research material. In this system, each participating country produces one-page publications abstracts that are circulated to the entire group through three centers: the British Road Research Laboratory (RRL), the French Central Laboratory for Bridges and Roads (LCPC), and the German Highway Research Board (FG). Abstracts of Canadian publications are prepared by the Canadian Good Roads Association and are sent both to HRIS and to the OECD system.

Special Committee on International Cooperative Activities

From its early years the Board has had ties with other countries in its information services. These ties have increased through the years both directly and through the Office of the Foreign Secretary of the National Academy of Sciences.

In 1925, in his address to the Annual Meeting of the Board, Thomas H. MacDonald said:

> Science extends beyond the borders of this country. It extends beyond the officialdom, beyond the universities, and beyond the commerce of this country. It extends beyond the borders of all countries. It is international, and this group must recognize that their labors are not only of great benefit nationally. I am more than ever convinced that they are of the greatest import from the international standpoint.

In the same year (1925) the first foreign contact man was appointed to provide liaison with the University of Mexico at Mexico City.

In 1926 the Highway Research Board joined the Permanent International Association of Road Congresses in Paris. It retained this affiliation through 1931, then joined again in 1967. In 1931 the Executive Committee "gave consideration to distributing a considerable number of HRB *Proceedings* abroad—in particular to delegates to the sixth PIARC."

In 1941 Director Crum attended the Pan-American Highway Congress in Mexico City. Upon his return the Executive Committee took note of the opportunity to extend informational service into the neighboring Pan-American states and authorized Mr. Crum to explore the possibility for arranging translations of pertinent publications into Spanish for distribution in Mexico and the Central and South American countries.

In 1959 the Board, acting jointly with the American Society of Civil Engineers and operating with a grant from the National Science Foundation, arranged for exchange visits with Soviet soils engineers.

In 1961 the Executive Committee authorized the Director to explore possibilities of conducting a worldwide census of highway research. The Board tried to enlist industry membership to provide funds for this and other expanded services. Industry interest in supporting the Board was limited at that time, and the proposed expansion was curtailed.

In his address marking the centennial of the National Academy of Sciences on October 22, 1963, President John F. Kennedy said: "I can imagine nothing more unwise than to hoard our knowledge and not disseminate it throughout the globe. . . ."

In 1965 M. Earl Campbell visited the Philippines, Thailand, India, Iran, and Lebanon in order to determine technological needs in highway work and also to determine the translation needs in highway literature—in particular the literature of the Highway Research Board. Later he went to Brazil to address the First Annual Meeting of the Highway Research Board of Brazil at Rio de Janeiro.

In 1966 Director D. Grant Mickle was invited to Australia to exchange experiences, and from there he went on to pay courtesy calls to several Asian countries.

In 1967 the Board again (after 36 years) affiliated with PIARC. Director William N. Carey, Jr., and Wilbur S. Smith, Chairman of the Special Committee on International Cooperative Activities, attended the PIARC meeting in Tokyo in November. HRB Chairman E. G. Wetzel attended the Tenth Pan-American Congress in Montevideo, Uruguay, in December 1967.

Attendance at the Annual Meetings from foreign countries has been increasing year by year. Some 30 countries now send about 200 representatives—some to present papers, some as members of HRB committees. There are affiliates in 73 countries.

The Board established a Special Committee on International Cooperative Activities in 1965. Wilbur S. Smith was appointed Chairman, and M. Earl Campbell served as Secretary until he left the Board in September 1967. Robert O. Swain took over the Chairmanship in 1970. R. C. Edgerton is Secretary.

This committee handles papers from abroad, arranges for appointments of foreign contact men, solicits articles on research organizations and research administration in other countries, handles inquiries from abroad, arranges for translations from and into foreign languages, and provides an annual program of papers relating to research in other countries. It serves uniquely and distinctively among many organizations in the international field.

Evolution of Technological Structure

The first part of this chapter dealt with the Board's administration of its overall program and its in-house staff engagement as well as administration of research projects and programs. This part deals with the volunteer effort—the involvement of more than 2,000 people (administrators, engineers, researchers, and educators) in committee work for the advancement of transportation and its environment.

The Research Program

In October 1919 three research subcommittees were established in the Committee on Highway Research, newly created as a unit of the Division of Engineering of the National Academy of Sciences–National Research Council. These were the Committee on Highway Economic Theory, the Committee on Structural Design of Roads, and the Committee on Character and Use of Road Materials.

A year later, when the Committee on Highway Research became the Advisory Board on Highway Research, these committees were included in the reorganization without change. Provision was made in the reorganization meeting for five additional committees. By 1923 seven committees had been set up in name, but only five functioned at that time. In addition to the three already named, a Committee on Maintenance and a Committee on Traffic were established.

On January 1, 1925, when the name was changed to the Highway Research Board, the five committees then functioning continued. Only one committee was added—the Committee on Properties of Bituminous Materials. By 1929, however, with a proliferation of activities, it became desirable to organize the actual and proposed activities into subject areas (sections) within these groups of related activities. This was accomplished in 1930, and at that time 33 committees were distributed among five sections:

1. Highway Transportation Economics (two committees);
2. Highway Design (seven committees);
3. Materials and Construction (eight committees);
4. Maintenance (five committees); and
5. Traffic (11 committees).

Five years later, with still greater activities and demands for further increases, the Board again reorganized. The sections were reorganized as departments with project and standing committees. Seven departments were created with an umbrella Advisory Committee. There is no record of the Advisory Committee's functions or activities.

At that time (1935) the Department of Soils and Investigations was established and became a vital unit of the Board; 11 committees came into being with its establishment. The seven departments now contained 36 committees.

With the expansion of departmental activities and the continuing increase in committees, there were internal reorganizations within the departments. In 1938 the Department of Traffic and Safety reorganized the scope of its interests and set up 10 new committees. At the end of 1941 it was resolved into a Department of Traffic and Operations with a new chairman and eight distinctive committees.

About the same time there was a review and some reorganization in the Departments of Design, Materials and Construction, and Soils and Investigations. Special committees reporting to the Director also began to appear.

In 1944 the Department of Economics, Finance and Administration was established by merging the Department of Finance and Administration with the Department of Highway Transportation Economics.

As years went by and activities and committees increased, departments instituted divisions. The Materials and Construction Department established four divisions in 1947, each with its own chairman. These divisions were Bituminous, Concrete, Construction Practices, and General.

In 1960 the Department of Economics, Finance and Administration instituted divisions, but dissolved them and returned to the simpler structure in 1967 because it had become apparent that coordination was difficult with the separation into divisions.

In 1963 the Department of Soils and Investigations was reorganized with a reorientation in scope, changing its name to Department of Soils, Geology and Foundations.

In 1964 two special committees were given department status—Department of Urban Transportation Studies (six committees) and Department of Legal Studies (five committees).

As of July 1, 1968, there were eight departments with 109 committees. Seven special committees were also functioning. The purpose and functions of the departments and departmental committees are summarized in Appendix I.

On February 1, 1970, a new consolidation went into effect. The eight existing departments were merged into one division (Division A—Regular Technical Activities) consisting of three groups. (The accompanying organization charts show the new group structure of the Board.) Group 1 is composed of the former Departments of Economics, Finance and Administration, Urban Transporation Planning, and Legal Studies. Group 2 is composed of the former Departments of Design, Materials and Construction, and Soils, Geology and Foundations. Group 3 is composed of the former Departments of Maintenance and Traffic and Operations. Special committees were assigned to the appropriate groups.

The evolution of the technological structure reflects two opposing forces—division and coordination. As growth continues, increasing the number of units to be administered, the usual process is to create subunits and then further subunits for purposes of housekeeping and distribution of management responsibilities. But as divisions are created the subunits become compartmentalized and isolated; the research tree becomes a host of broken branches, almost impossible to keep viable as a component. In other words, communication becomes so difficult that coordination is almost impossible.

Hence, it is imperative to start with a realistic framework for research built on transportation objectives and divide this into "natural" components for administration, and then to effect feasible communication links among the components to prevent compartmentalization and isolation with subsequent departure from objectives.

To try to give all of the names of committee members from 1920 to the present would call for a listing of more than 4,500 names. These names are included in other publications of the Board, including the *Proceedings* and *Yearbooks*. Appendix J shows the evolution of the structure from committee to department and thence to group.

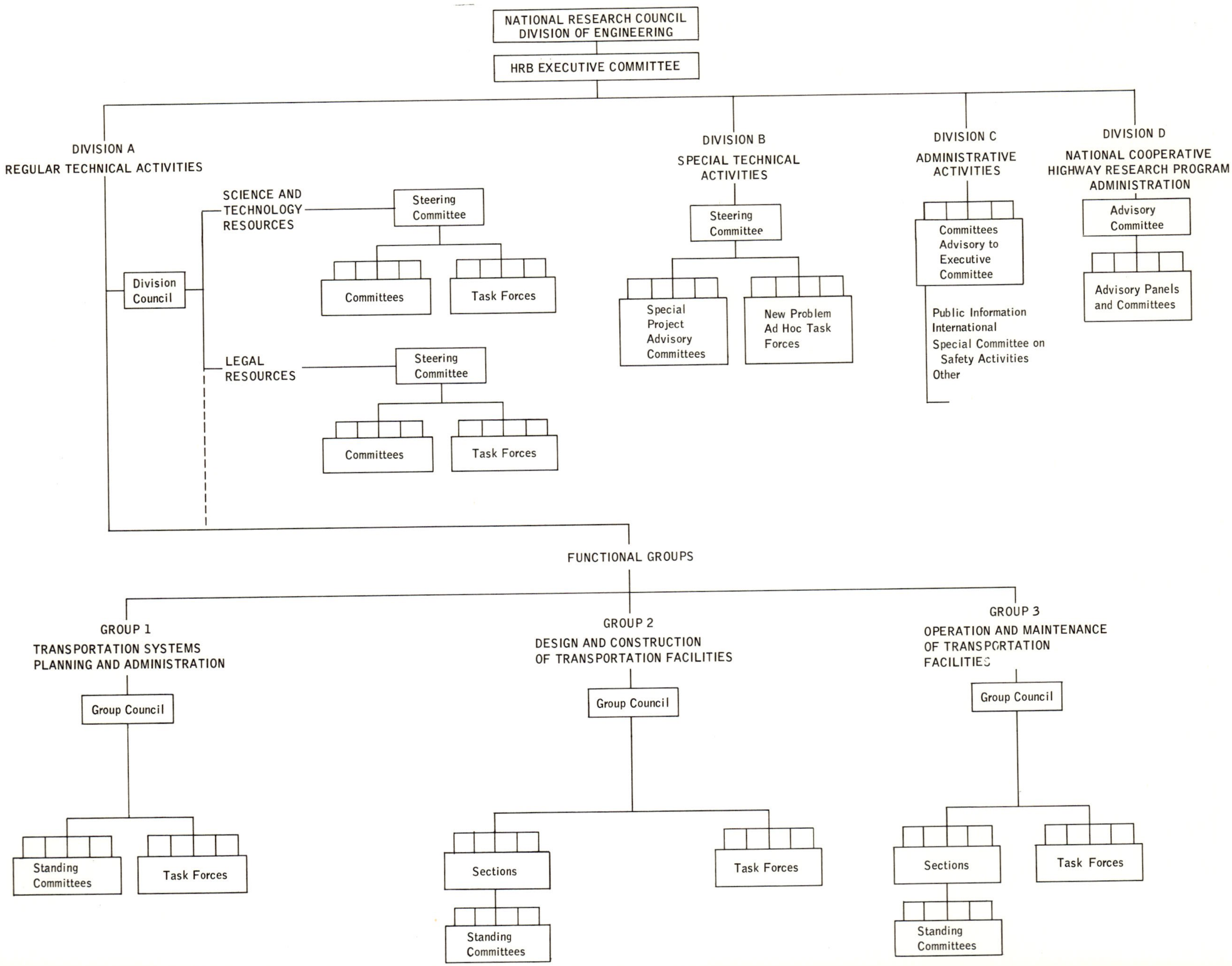
NATIONAL RESEARCH COUNCIL
DIVISION OF ENGINEERING
HRB EXECUTIVE COMMITTEE
DIVISION A
REGULAR TECHNICAL ACTIVITIES
Division Council
SCIENCE AND TECHNOLOGY RESOURCES
Steering Committee
Committees
Task Forces
LEGAL RESOURCES
Steering Committee
Committees
Task Forces
DIVISION B
SPECIAL TECHNICAL ACTIVITIES
Steering Committee
Special Project Advisory Committees
New Problem Ad Hoc Task Forces
DIVISION C
ADMINISTRATIVE ACTIVITIES
Committees Advisory to Executive Committee
Public Information
International
Special Committee on Safety Activities
Other
DIVISION D
NATIONAL COOPERATIVE HIGHWAY RESEARCH PROGRAM ADMINISTRATION
Advisory Committee
Advisory Panels and Committees
FUNCTIONAL GROUPS
GROUP 1
TRANSPORTATION SYSTEMS PLANNING AND ADMINISTRATION
Group Council
Standing Committees
Task Forces
GROUP 2
DESIGN AND CONSTRUCTION OF TRANSPORTATION FACILITIES
Group Council
Sections
Standing Committees
Task Forces
GROUP 3
OPERATION AND MAINTENANCE OF TRANSPORTATION FACILITIES
Group Council
Sections
Standing Committees
Task Forces

HIGHWAY RESEARCH BOARD

GROUP 1

TRANSPORTATION SYSTEMS PLANNING AND ADMINISTRATION

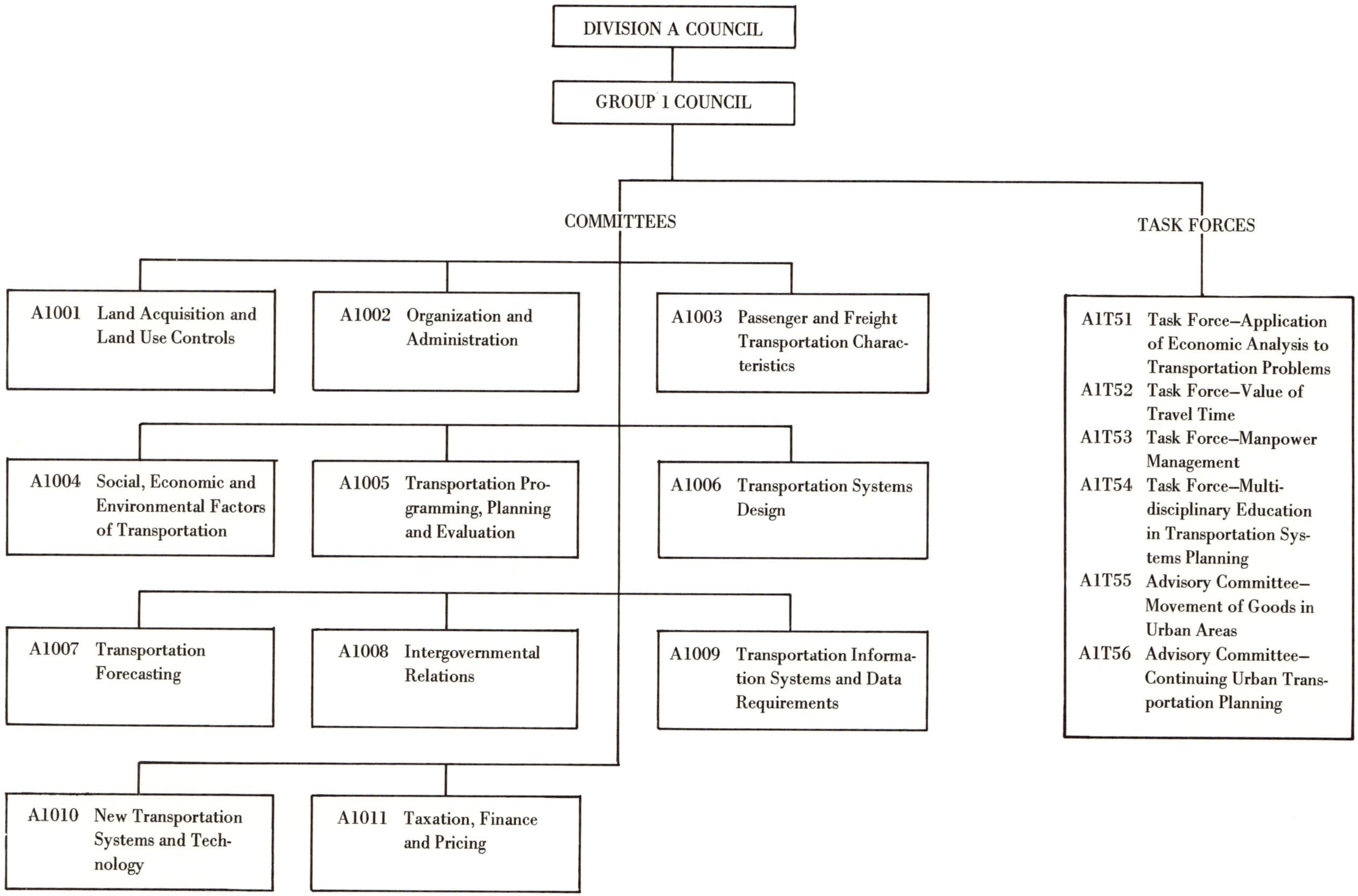

HIGHWAY RESEARCH BOARD

GROUP 2

DESIGN AND CONSTRUCTION OF TRANSPORTATION FACILITIES

DIVISION A COUNCIL

GROUP 2 COUNCIL

GENERAL DESIGN SECTION

Committees

- A2A01 Photogrammetry and Aerial Surveys
- A2A02 Geometric Highway Design
- A2A03 Surface Drainage of Highways
- A2A04 Traffic Safety Barriers and Sign, Signal and Lighting Supports
- A2A05 Roadside Development
- A2A06 Subsurface Structures Design
- A2A07 Shoulder Design

PAVEMENT SECTION

Committees

- A2B01 Rigid Pavement Design
- A2B02 Flexible Pavement Design
- A2B03 Design of Composite Pavements and Structural Overlays
- A2B04 Surface Properties—Vehicle Interaction
- A2B05 Pavement Condition Evaluation
- A2B06 Theory of Pavement Design

BRIDGE SECTION

Committees

- A2C01 Bridge Design
- A2C02 Steel Superstructures
- A2C03 Concrete Superstructures
- A2C04 Substructures, Retaining Walls and Foundations
- A2C05 Dynamics and Field Testing of Bridges

BITUMINOUS SECTION

Committees

- A2D01 Characteristics of Bituminous Materials
- A2D02 Effects of Natural Elements and Chemicals on Bitumen-Aggregate Combinations
- A2D03 Bituminous Surface Treatments
- A2D04 Mechanical Properties of Bituminous Paving Mixtures
- A2D05 Characteristics of Aggregates and Fillers for Bituminous Construction
- A2D06 Bituminous Aggregate Bases
- A2D07 General Asphalt Problems

CONCRETE SECTION

Committees

- A2E01 Performance of Concrete—Physical Aspects
- A2E02 Performance of Concrete—Chemical Aspects
- A2E03 Mechanical Properties of Concrete
- A2E04 Curing of Concrete
- A2E05 Chemical Additions and Admixtures for Concrete
- A2E06 Basic Research Pertaining to Portland Cement and Concrete

TASK FORCES

As Required

- A2T51 Task Force—Automation Systems for Highway Organizations
- A2T52 Task Force—Transporting Abnormally Heavy Loads on Highways
- A2T53 Task Force—Structural Design of Pavement Systems
- A2T54 Task Force—Notch Toughness of Steels
- A2T55 Advisory Committee—Structural Design of Asphalt Concrete Pavement Systems Workshop

CONSTRUCTION SECTION

Committees

- A2F01 Rigid Pavement Construction
- A2F02 Flexible Pavement Construction
- A2F03 Earthwork Construction
- A2F04 Construction of Bridges and Structures
- A2F05 Construction Management
- A2F06 Highway Equipment
- A2F07 Connection Procedures—Metal Structures

GENERAL MATERIALS SECTION

Committees

- A2G01 Mineral Aggregates
- A2G02 Coatings, Signing and Marking Materials
- A2G03 Sealants and Fillers for Joints and Cracks
- A2G04 Adhesives, Bonding Agents and Their Uses
- A2G05 Corrosion

EVALUATIONS, SYSTEMS AND PROCEDURES SECTION

Committees

- A2H01 Nuclear Principles and Applications
- A2H02 Quality Assurance and Acceptance Procedures

COMPACTION AND STABILIZATION SECTION

Committees

- A2J01 Compaction
- A2J02 Chloride Stabilization
- A2J03 Lime and Lime-Fly Ash Stabilization
- A2J04 Soil-Portland Cement Stabilization
- A2J05 Soil-Bituminous Stabilization
- A2J06 Chemical Stabilization of Foundations

SOIL MECHANICS SECTION

Committees

- A2K01 Strength and Deformation Characteristics of Pavement Sections
- A2K02 Embankments and Earth Slopes
- A2K03 Foundations of Bridges and Other Structures
- A2K04 Subsurface Soil-Structure Interaction
- A2K05 Mechanics of Earth Masses and Layered Systems
- A2K06 Subsurface Drainage

SOIL AND ROCK PROPERTIES AND GEOLOGY SECTION

Committees

- A2L01 Exploration and Classification of Earth Materials
- A2L02 Soil and Rock Properties
- A2L03 Physicochemical Phenomena in Soils
- A2L04 Frost Action
- A2L05 Engineering Geology
- A2L06 Environmental Factors Except Frost

HIGHWAY RESEARCH BOARD

GROUP 3

OPERATION AND MAINTENANCE OF TRANSPORTATION FACILITIES

DIVISION A COUNCIL

GROUP 3 COUNCIL

CONTROL AND COMMUNICATIONS SECTION

Committees

- A3A01 Communications
- A3A02 Traffic Control Devices
- A3A03 Motorist Information Systems
- A3A04 Visibility
- A3A05 Highway-Railroad Grade Crossings

FACILITY USERS SECTION

Committees

- A3B01 Vehicle Characteristics
- A3B02 Road User Characteristics
- A3B03 Driver Education
- A3B04 Pedestrians
- A3B05 Motorist Services
- A3B06 Simulation of Driving Task

MANAGEMENT SECTION

Committees

- A3C01 Maintenance and Operations Management
- A3C02 Maintenance and Operations Costs
- A3C03 Maintenance and Operations Personnel
- A3C04 Traffic Law Enforcement
- A3C05 Vehicle Inspection and Regulation

MAINTENANCE OF FACILITIES SECTION

Committees

- A3D01 Pavement Maintenance
- A3D02 Structures Maintenance
- A3D03 Roadside Maintenance
- A3D04 Maintenance Equipment
- A3D05 Maintenance of Traffic Control Devices
- A3D51 Task Force–Effect of Studded Tires

OPERATION OF FACILITIES SECTION

Committees

- A3E01 Parking and Terminals
- A3E02 Operational Effects of Geometrics
- A3E03 Freeway Operations
- A3E04 Snow and Ice Control
- A3E05 Maintenance Operations
- A3E06 Transportation of Hazardous Materials

SYSTEMS PERFORMANCE SECTION

Committees

- A3F01 Highway Capacity and Quality of Service
- A3F02 Traffic Flow Theory and Characteristics
- A3F03 Traffic Records
- A3F04 Effectiveness of Operational Measures

HIGHWAY RESEARCH BOARD

LEGAL RESOURCES GROUP

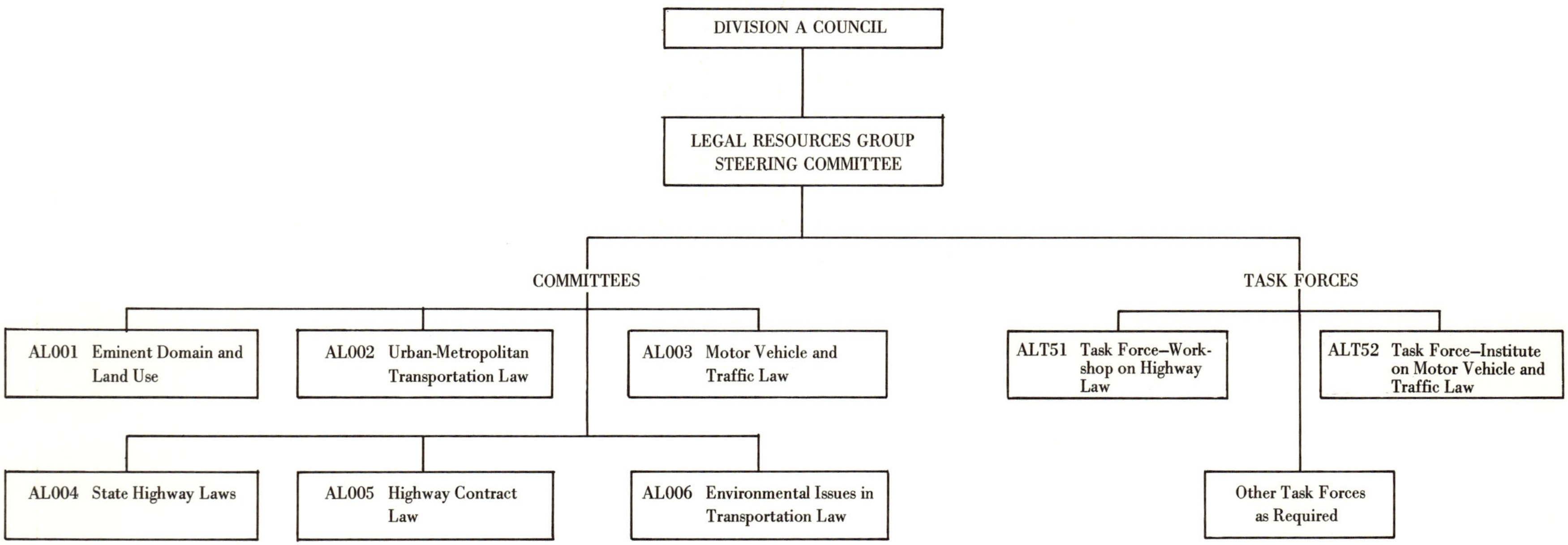

VETERAN STAFF MEMBERS OF THE HIGHWAY RESEARCH BOARD

During the mid-Forties the staff of the Highway Research Board was greatly expanded, largely due to the formation of the Research Correlation Service and the resultant need for engineers. Among the staff members employed during this period were Wainwright Bridges, who became Administrative Assistant to the Director in 1945; A. Walton Johnson, Franklin N. Wray, and Elmer M. Ward, who were employed in 1946 as Engineer of Soils and Foundations, Engineer of Design, and Engineer of Equipment and Maintenance, respectively; and E.W. Harris, who joined the Board in 1946 as an assistant in the business office and later became Assistant to the Executive Director. M. Earl Campbell was employed as Engineer of Traffic and Operations in 1947. Later additions to the engineering staff included Kenneth McWane in 1952 and Ray E. Bollen in 1955.

Wainwright Bridges

E. W. Harris

Ray E. Bollen

A. Walton Johnson

Elmer M. Ward

M. Earl Campbell

Kenneth McWane

Franklin N. Wray

ASSISTANT DIRECTORS OF THE HIGHWAY RESEARCH BOARD, 1970

Roy C. Edgerton

Krieger W. Henderson, Jr.

James C. Allen

Paul E. Irick

The Annual Meetings of the Highway Research Board traditionally attract delegates and speakers from overseas as well as from the United States. Sir William Glanville, Director of Britain's Road Research Laboratory, is shown addressing a session at a recent meeting, while C.D. Curtiss, Commissioner of the U.S. Bureau of Public Roads and past Chairman of the Executive Committee of HRB, looks on. Both Sir William and Mr. Curtiss have since retired from their respective posts.

5

Relationship Between the State Highway Departments and the HRB

The close relationship between the Highway Research Board and the state highway departments as represented by the American Association of State Highway Officials began on November 11, 1920. Then, upon invitation from Chairman C. A. Adams of the Division of Engineering of the National Research Council, a group of representatives from 10 national associations, societies, and agencies and six universities and colleges met in the Engineering Societies Building in New York City and organized the Board. C. J. Bennett, State Highway Commissioner of Connecticut, represented AASHO at the meeting. Thus began the Board's association with AASHO, which has been continuous for 50 years and has substantially expanded the Board's technical relationships, both with AASHO and with the individual state highway departments. This brief history reviews the highlights of these technical relationships.

In July 1921, only 8 months after the Board was organized, Alfred D. Flinn, Vice Chairman of the Division of Engineering, reported: "Ten state highway departments and 44 universities have indicated interest in the work of your Board and promised cooperation." On December 8, 1921, W. K. Hatt, who later became Director of the Board, addressed the Seventh Annual Meeting of AASHO at Omaha, Nebraska, on the subject, "The Coordination of Highway Research." At this same AASHO meeting the following encouraging resolution was adopted:

Be it resolved that the American Association of State Highway Officials welcomes the establishment by the National Research Council of the Advisory Board on Highway Research, and that the Association pledges its support to the Advisory Board in this much-needed effort to stimulate and coordinate highway research.

This support in the early pioneering years represented only a relatively small amount of money. In fact, the only financial support received from the states in the early years was a $1,000 per year grant for 2 years (1922 and 1923) from the Connecticut State Highway Department. The chief difficulty in obtaining financial support was the legal inability of the states to enter into cooperative agreements. Several state highway departments offered to cooperate with the Board's committees, provided the funds they offered would be expended only in their respective states.

In these early pioneer years, the spirit of sincere cooperation by the chief administrators of the state highway departments was manifested in their willingness to (a) have key members of their departments serve on or chair the all-important technical committees of the Board; (b) allow qualified personnel to attend and in many cases to present papers on research findings at Annual Meetings of the Board; (c) supply upon request fundamental information on general research matters; and (d) render helpful and valuable assistance to staff members of the Board assigned to make special studies that involve personal visits to the state highway departments.

The record shows that the members of the five committees of the Board in 1922 included three chairmen and that 16 of the 48 committee members were from state highway departments. The Bureau of Public Roads was represented by seven members including one committee chairman, and there were 16 members from 15 universities and colleges, one of whom was a committee chairman.

Of the total attendance of 273 at the Annual Meeting in 1924, there were 68 representatives from 18 states; 10 of these representatives were listed on the general program.

Another example of the willingness of the state highway departments in furnishing information on general research matters is exemplified in the first census of highway engineering research made and published by the Board in 1922. This material was compiled from information obtained by a questionnaire. The resulting bulletin, *Highway Research Projects in the United States,* contained a census of 479 current or completed research projects in highway engineering and highway transport. Highway departments in 40 states responded by submitting detailed information on 179 projects—about 40 percent of the total.

During the 1920's the Board administered three special studies financed by industry. These were Investigation on Economic Value of Steel Reinforcement in Concrete, Low-Cost Road Investigation, and Culvert Investigation. Special staff personnel were employed by the Board to supervise each study. The cooperative assistance accorded the staff by the engineers of the state highway departments in these studies was outstanding. In the study on steel reinforcement conducted in 1924 and 1925, an inspection was made of 5,500 miles of plain and reinforced concrete roads in 21 states. The following statement, taken from the report, emphasized the important role of the state highway engineers in this study: "The state highway department personnel furnished reports in such detail that if we had to gather the information ourselves, the investigation would have taken several years."

In the study on low-cost roads, valuable assistance was rendered the Board's staff engineer from conferences with engineers from 25 highway departments. In the culvert study, conferences and interviews with highway engineers from 14 highway departments were of great assistance in the identification and classification of local conditions during the examination of 516 culverts.

In all of these studies, the state highway department engineers were most willing and liberal not only in supplying valuable information but also in providing field transportation and engineering personnel assistance to the Board's staff members. These contributions were just as important to the success of the studies as the finances provided by industry.

The relationship between the two organizations was strengthened in 1924 by the appointment of active representatives on the Board from the state highway departments. These representatives, or contact men, made it possible for the state highway departments to secure immediate benefit from the activities of the Board because they received, without cost, a copy of each publication of the Board as soon as it was printed. This arrangement also made it possible for the Board to keep in touch with the problems of the states in the administration, design, construction, operation, and maintenance of highways. This mutually useful arrangement has endured and grown over the years, providing a two-way flow of information.

In 1931 the Board created a special committee to study and report on cooperative arrangements with the American Association of State Highway Officials. The following statements are taken from the committee's 1932 report:

> Your committee is pleased to report that informal arrangements have been made for facilitating the progress of research work that is of mutual interest to the American Association of State Highway Officials and the Board.
>
> We are in agreement with the officers of the Association in believing that no

formal organizations, such as permanent committees on cooperation, are necessary or advisable.

We are agreed that cooperative projects between the Association and the Board, acting through its various committees, should be organized whenever a need for a specific piece of work of mutual interest arises, and upon which the prospects of accomplishment are reasonably good. The project should be subject to the approval of the proper officers of each organization.

The first joint cooperative project established between the Board and AASHO under this arrangement was "A Study of the Laws, Funds, Organizations, and Technical Practices Relating to Roadside Development." This study was made in 1932 by a joint committee consisting of four members appointed by the Association and three members appointed by the Board. Sixteen state highway departments having definite policies for such work furnished the detailed information for the first report on this study.

Also in 1931 the Board accepted joint sponsorship with AASHO and the American Road Builders' Association for the award each year of a bronze plaque to someone performing notable service in the highway field, in honor of George S. Bartlett. A fund sufficient to cover all expenses of this award had been contributed by friends of Mr. Bartlett, "to perpetuate the spirit of friendship and helpfulness which he brought into his work in the highway field."

Another indication of the cooperation between the Board and AASHO was the approval by the Executive Committees of the two agencies of a joint arrangement for conducting, maintaining, and publishing periodic highway research censuses. This was inaugurated in 1936 when the Association appointed the Director of the Board to be Secretary of its Committee on Research Activities. This arrangement still continues. The Association and the Board also shared jointly the expenses of publishing the first two publications issued under this arrangement.

During the 1930's two special publications were issued in which the cooperative support of the state highway departments again was a major factor in their success: *Use of Calcium Chloride as a Dust Palliative* and *Curing of Concrete Pavements*. The actual financial support for these projects was furnished by several industrial agencies.

In the study of dust palliatives, the generous cooperation of three state highway departments included the use of several miles of existing roads as test sections for securing data on the dispersion of calcium chloride by chemical action, rainfall, and maintenance manipulation. The laboratories of the departments also made numerous physical and chemical tests during the time the test roads were under observation. These research projects were administered by a special committee of the Board through a staff engineer.

The curing study was also made possible through the cooperation and support of 18 state highway departments and six industrial agencies. Again the latter agencies furnished the financial assistance, most of which was used to pay the salary of the special investigator and the expenses of the committee members to attend meetings. But the state highway department administrators also made very important contributions, such as furnishing engineers and field transportation for inspection of surface conditions of approximately 2,000 miles of concrete road, allowing department engineers to serve on the special curing committee, and turning over the curing research reports of the departments to the committee for review, analysis, and correlation.

In 1943 a special committee, appointed by AASHO and working in cooperation with the Board, prepared a plan for the Research Correlation Service. This plan was approved in the following resolution adopted at the 28th Annual Meeting of the Association at Chicago on December 3, 1943:

> WHEREAS, it is apparent that all highway administrations are faced with the imminent prospect of vastly increased highway rehabilitation and development; and
>
> WHEREAS, a greatly enlarged program of research and correlation and utilization will be necessary in order to provide for doing the work with maximum speed, economy, and efficiency; and
>
> WHEREAS, present facilities do not permit current correlation to the extent that results of research can be exchanged and used prior to completion and a considerable lag in practical use results; and
>
> WHEREAS, these matters have been considered by a special joint committee appointed by the Executive Committee of the American Association of State Highway Officials and the Highway Research Board, which has submitted a unanimous report calling attention to the urgent need for increased correlation facilities and has submitted a recommended procedure and budgets for achieving the desired results; and
>
> WHEREAS, it is to the public interest that additional funds be provided from any available public sources for the required increase in research activities, correlation, and provision for their practical use;
>
> THEREFORE, be it
>
> RESOLVED, that the American Association of State Highway Officials, recognizing the need for an increased program of correlation and utilization of research, directs its Executive Committee to promote such an expanded program as in its judgment is necessary, and copies of these resolutions be sent to the Highway Research Board.

Congress also recognized that research was an essential feature of the continued development of highway facilities by providing in the 1944 Federal-Aid Act an expansion of the areas in which the research funds could be expended. The interpretation of this Act by the Public Roads Administration gave the following statement in respect to research:

> One and one-half per cent of the total amount apportioned to any State under Section 4 of the Act will be reserved for purposes of engineering and economic investigations and research necessary in connection therewith. States will be requested to match in the statutory ratio, the Federal funds used for these purposes.
>
> The State Highway Departments will be advised that the use of small amounts of the apportioned funds for the purpose of contribution to the expense of maintaining a highway research correlation service by the Highway Research Board will be approved as within the purposes of Section 8, providing the Federal funds so used are matched with State funds in the statutory ratio.

Under this plan, which was inaugurated in 1945, the Highway Research Board acts as the agent for AASHO in organizing and administering the Research Correlation Service by means of field engineers, committee work, and publications. This service is financed on a yearly subscription basis by the state highway departments, the Bureau of Public Roads, and other funding.

To implement RCS the Board has a staff of technical men, each of whom has specialized in one or more of the branches of highway technology represented by the departments and special committees of the Board. These staff members spend a substantial part of their time making periodic visits to the state highway departments and other organizations engaged in highway research, where they discuss problems of operations and research with administrators and researchers and thus acquire first-hand information on developments in the nationwide highway research efforts. Through these field contacts by the technical staff, an effective linkage is created among the state highway departments, federal agencies, colleges, universities, and industrial organizations engaged in highway research.

In addition to liaison provided by the staff engineers, other special services are provided, such as help in formulating research projects, preparation of special bibliographies, search for specific library information, compilation of regional practices or procedures relating to special problems, and preparation of lectures for conferences or schools. These staff engineers also provide their respective departments and committees with technical assistance.

This service has proved most rewarding to both the state highway departments and to the Board, and has made the relationship between them both closer and more meaningful. The upper of the two accompanying figures shows the stimulating effect that the increase in the Board's activities, made possible by the states' contributions to the Research Correlation Service, has

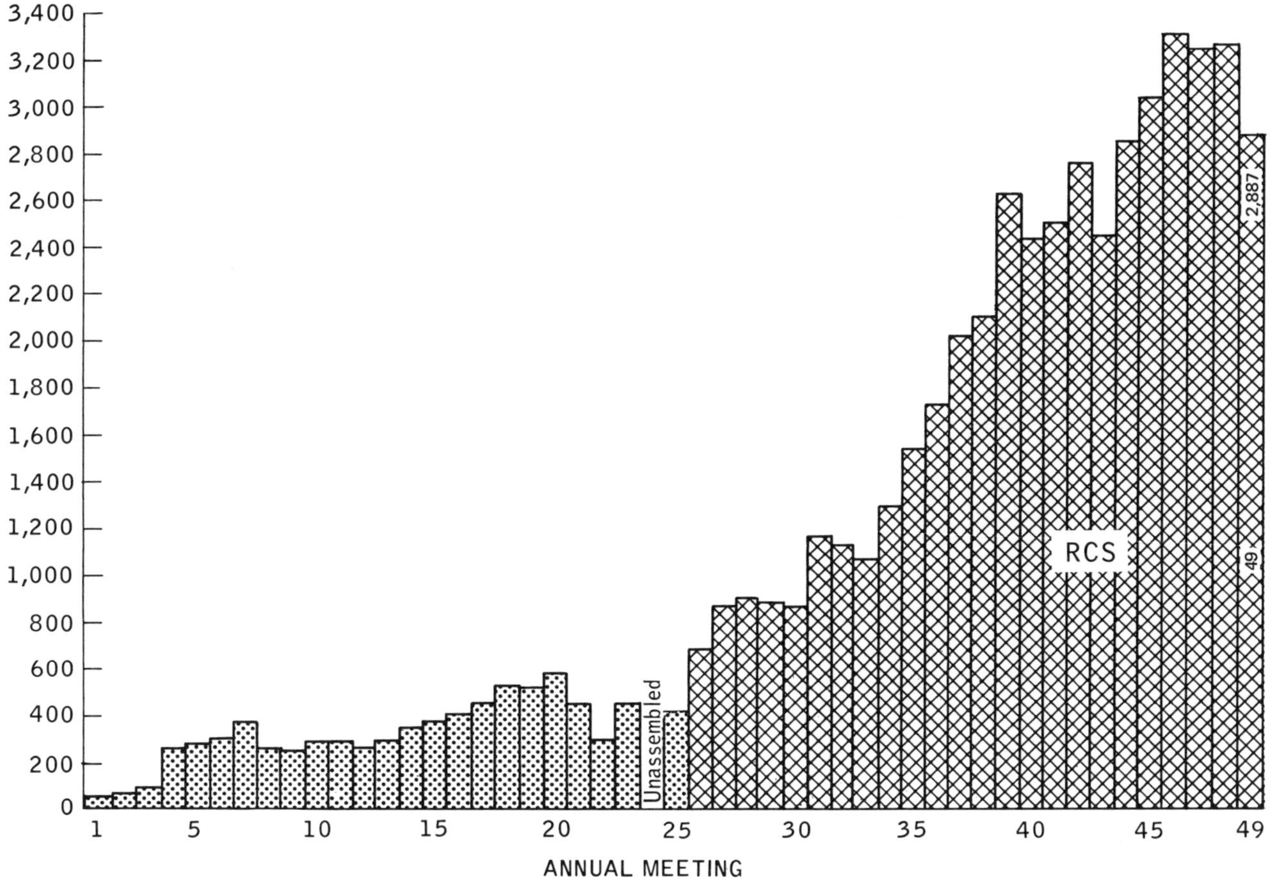

Attendance at Annual Meetings.

Number of pages published during fiscal year.

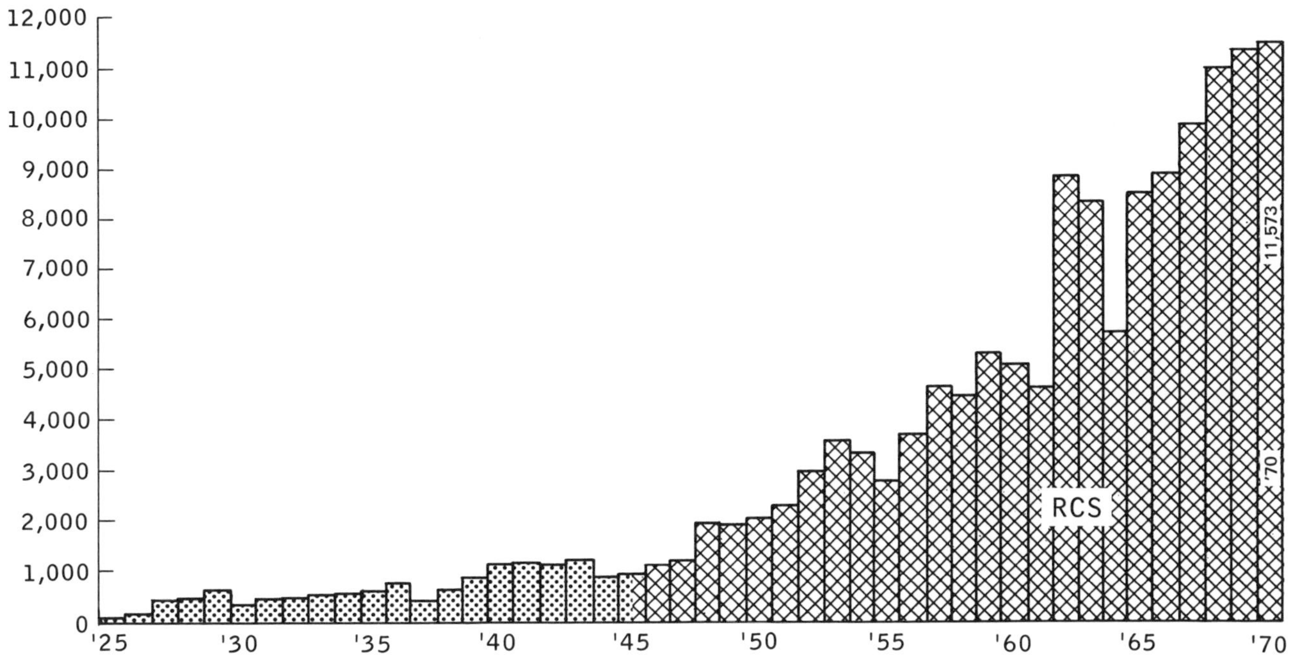

had on research and the important part this increase has played in attracting technical personnel to the Annual Meetings. The lower figure shows a striking relationship between the increases in the states' contributions to the service and the corresponding increase in the Board's activities, as measured by the number of pages of technical matter published before and after inauguration of the RCS. This valuable research material is distributed as soon as it is published to approximately 1,600 key highway engineers in all the state highway departments. Prompt and widespread dissemination of new findings is considered essential to highway engineering progress, for it is the practical highway engineers who are in a position to appreciate, interpret, and apply the findings, most of which shed new light on important engineering, economic, and social problems.

The two figures show the overall success of the Research Correlation Service to both the state highway departments and the Highway Research Board. The state highway departments were benefited mainly in that many of their problems were subjected to fresh scientific studies that produced the needed new knowledge for greater practical utilization of the findings. The Board in turn was benefited by the increased financial support that enabled it to broaden the scope of its activities and usefulness. (The lower figure also shows that the users received increased services proportionate with the increase in financial support given the providers.)

The faith and confidence that the American Association of State Highway Officials had in this new service is expressed in the following resolution adopted at the Association's 31st Annual Meeting on February 1, 1946, less than 1 year after the service was inaugurated:

> WHEREAS, the American Association of State Highway Officials notes with approval that arrangements have been completed between State, Territorial, and District of Columbia Highway Departments and the Highway Research Board of NRC–NAS, with the approval of the Public Roads Administration for the inauguration of the Highway Research Correlation Service:
>
> THEREFORE, be it
>
> RESOLVED: That the Association appreciates the cooperation of the Highway Research Board in the establishment of this fine Service; and be it further
>
> RESOLVED: That this Association recommends and urges full use and continuation of the Service.

By unanimous vote of the member organizations of the Board, the Bylaws were amended in 1945 to permit the Executive Secretary of the American Association of State Highway Officials to serve as an ex officio member of the Executive Committee of the Board. Hal H. Hale filled this position until 1955. He was succeeded by Alfred E. Johnson, who is still

Executive Secretary of AASHO and an ex officio member of the Executive Committee in 1970.

In 1946 the Board and AASHO established a Joint Committee on Maintenance Personnel. This joint committee has been presenting reports at the Annual Meetings of the Association and Board ever since. All the state highway departments have furnished information from which these highly regarded annual reports are compiled. They contain tabulations showing comparative current wages, salaries, sick leave, vacation, and retirement benefits for maintenance personnel in the state highway departments. Over the years these reports have been instrumental in establishing equitable wage scales and in improving the general personnel conditions in the maintenance departments.

A procedure for formulation and administration of research projects recommended by committees of the American Association of State Highway Officials to be financed by two or more states was approved by the Executive Committee on September 24, 1948, upon recommendation of both the Committee on Standards and the Research Activities Committee of AASHO. According to the procedure, a committee of AASHO desiring to recommend a research project for joint financing by two or more states, with or without use of federal-aid 1½ percent funds, should submit a project statement and estimates of cost to the Committee on Research Activities. The project statement should show the need for the research, the scope of the proposed work, and the states that might be expected to take part. The project statement would be referred for comment to the chairman of other appropriate committees of AASHO. After due consideration the Committee on Research Activities would transmit a report of its findings with recommendation for action to the Chairman of the Committee on Standards. If recommended by the Committee on Research Activities, the Chairman of the Committee on Standards would submit the proposed project to the consideration of the states that may be concerned in its financing and the Bureau of Public Roads, and ascertain which states are willing and able to participate. The Chairman of the Committee on Standards would then notify the Highway Research Board of the proposed project and of the interested agencies. Upon receipt of this notification, the Highway Research Board would offer its Research Correlation Service to the states concerned and to the BPR for the purpose of arranging, to the satisfaction of all contributing states, a scheme of joint financing. If desired by the states concerned, and upon approval of its Executive Committee, the Highway Research Board would undertake supervision and administration of the jointly arranged project, proceeding (a) to enter into agreements with the individual states for payment to it of contributions to the support of the work; (b) to appoint a project committee to plan and direct the work, the committee to include representatives of the Committee on Research Activities and of the committee sponsoring the project; (c) to arrange the necessary staff and proceed with the work; and (d) to prepare and publish reports of the project in accordance with the terms of its agreement and with the approval of the Committee on Research Activities, and to distribute such reports to all states and the Bureau of Public Roads. Approval of this procedure was an important and progressive step because it provided the basis for financing and carrying out many future joint research projects by the state highway departments, the Bureau of Public Roads, and the Highway Research Board.

In 1950 a joint committee of AASHO and the Board made the following report of a plan of action for cooperation between the two organizations:

Summary—The principal business of the committees of the Association is the production of Standards.

The principal business of the Board is research in all of its ramifications, including planning, promotion, evaluation of results, presentation of reports, and dissemination of information.

A major function of the Board is production of information needed for the formulation of Standards.

Although committees of the Association may properly engage in research activities, the Board has been retained by the Association to perform in this field and hence the committees of the Association can greatly extend their effectiveness by making use of the facilities of the Board.

Wasteful duplication of effort can be obviated by proper liaison among the operating committees of both organizations.

The relationship between the state highway departments and the Board has become increasingly close, mainly because of the progressive attitude of the state highway engineers in recognizing the value of joint research efforts on problems of common interest and their willingness to finance such projects under the general administration of the Board. In addition to several small projects, six major research projects have been jointly financed by the state highway departments and administered by the Board during the past decade and a half. These are discussed in Chapter 7.

Road Test One–MD was a field research project designed to determine the relative effects of 18,000-lb and 22,400-lb single-axle trucks and 32,000-lb and 44,000-lb tandem-axle trucks on a section of 24-ft wide (two 12-ft lanes) reinforced concrete pavement located on US 301 in southern Maryland.

Investigation of Wind Forces on Highway Bridges was a special cooperative research project to investigate, by means of wind tunnel studies, the effects of wind forces on bridge models of various designs.

The WASHO Road Test was conceived and sponsored by the Western Association of State Highway Officials and administered by the Highway Research Board. The full-scale field research test project was designed to determine the relative effects of 18,000- and 22,000-lb single-axle loads and 32,000- and 40,000-lb tandem-axle loads on flexible pavements with thicknesses ranging from 6 to 22 in.

The Highway Laws Study was requested by the Executive Committee of AASHO, and a Committee on Highway Laws was established in the Board's Department of Economics, Finance and Administration to supervise the project in 1952. The initial financing for the project was provided by grants from the Bureau of Public Roads and the Automotive Safety Foundation.

In 1960, upon recommendation of the AASHO Executive Committee, the state highway departments increased their annual contributions to the Research Correlation Service by 10 percent to ensure continuation of activity in this important field. This enabled the Board to employ a full-time counsel for legal research and a secretary, and to change the status of the Law Committee to that of a Special Committee. In June 1963 the Executive Committee of the Board, upon formal request by the Special Committee on Highway Laws, created a Department of Legal Studies. This arrangement allowed the scope of legal research and correlation to be enlarged and studied more effectively through the creation of appropriate and necessary subsidiary advisory committees. The AASHO Committee on Legal Affairs requested in 1962 that the Board organize and jointly sponsor a series of workshops for the state highway department legal counsels. These workshops, held annually, have been very successful and have attracted legal counsel from more than 30 highway departments at each meeting.

The AASHO Road Test (see Chapter 7) was conceived and sponsored by the American Association of State Highway Officials as a study of the performance of pavement structures of known thickness under moving loads of known magnitude and frequency. Portland cement concrete and asphaltic pavements, as well as certain types of bridges, were included in the test facility.

The National Cooperative Highway Research Program was initiated in 1962 as an objective national highway research program employing modern practices. This Program is supported on a continuing basis by funds from participating member states of the American Association of State Highway Officials and receives the full cooperation and support of the Federal Highway Administration, United States Department of Transportation.

A full explanation of the organization and operation of the NCHRP may be found in Chapter 7.

Summary

There is little doubt that the cooperative relationship among the state highway departments, the Bureau of Public Roads, and the Highway Research Board has been a significant factor in the advancement of highway transportation to its present high level. The highway research activities nurtured and supported by the states through the years have yielded new knowledge of vital importance to the welfare of the nation. The success of this continuous cooperative relationship has been due mainly to the abilities, ambitions, and incentives of the people involved. Their complementary skills and capabilities and their recognition of each other's viewpoints have produced this expanding cooperative activity with mutual advantages to all participating agencies.

From 1944 through 1954, the time of the development of the Research Correlation Service and the Maryland and WASHO Road Tests, Hal H. Hale served as Executive Secretary of AASHO. Since 1955, during the AASHO Road Test and the formation of NCHRP, Alfred E. Johnson was Executive Secretary (later Executive Director). The Board owes much to these two men. It was their understanding of research on the one hand and highway department administration on the other that frequently bridged the gap between researcher and administrator. They helped the AASHO-HRB relationship flourish.

6

Relationship Between the Bureau of Public Roads and the HRB

Bureau of Public Roads engineers were cooperating in the activities of the Division of Engineering of the National Research Council about a year before the Board was created. This began on October 8, 1919, when Dean Anson Marston and Professor A. N. Talbot, the representatives of the American Society of Civil Engineers and the Western Society of Engineers in the NRC Division of Engineering, met in Chicago with the following engineers of the Bureau of Public Roads: T. H. MacDonald, Chief of the Bureau; A. T. Goldbeck, Engineer of Tests and Investigations; T. W. Allen, Chief, Division of Control; and J. T. Voshell, District Engineer. The meeting was held to discuss the importance and necessity of the immediate inauguration of a national program for highway research. Clifford Older, President of the Mississippi Valley Conference of State Highway Departments, was also present. The outcome of the meeting was the creation by the Division of Engineering of the Committee on Highway Research on October 27, 1919.

The liaison with the Bureau of Public Roads was further strengthened when T. H. MacDonald, who had been appointed Chief of the Bureau on May 5, 1919, was added to the membership of the Division of Engineering on December 20, 1919.

The meeting of October 8, 1919, had aroused much interest in the need for organizing an active cooperative program of highway research; and during the interval of approximately a year before the Board was created, several

TABLE 1 Tentative Outline for a Comprehensive Program of Investigation of Problems Relating to the Design, Operation, and Economic Value of Highways

INVESTIGATIONS OF THE EFFECT ON ROAD CONSTRUCTION AND MAINTENANCE OF:	INVESTIGATIONS OF THE EFFECT ON THE ECONOMICS OF ROAD OPERATION OF:	INVESTIGATIONS OF THE EFFECT OF THE ECONOMIC VALUE OF HIGHWAYS TO A COMMUNITY OF:
1. Physical geographic conditions	*1. Road planning*	*1. Highway traffic*
Climate	Financing	Distribution of traffic
Topography	Environment	Geographical (relation to environment)
Geology	Farms	Seasonal (time of year)
2. Subgrade	Mines	Daily (day of week and month)
Character of soil	Timber lands	Hourly (hourly variation)
Moisture	Manufacturing centers	Character of traffic
Preparation	Other transportation systems	Passenger
3. Shoulders	Railways	Horse-drawn
Width	Ocean, rivers, canals	Motor
Character	Roadside treatment	Freight
Roadside vegetation, trees, etc.	*2. Grade*	Horse-drawn
4. Drainage	*3. Surface*	Motor
Bridges, culverts, etc.	*4. Alignment*	Weights per vehicle; per wheel; per inch of tread
5. Type of pavement	*5. Traffic movement*	Relation to other transportation systems
Materials	Regulations	Rail
Construction methods	Grade separations	Water
6. Loads imposed by traffic	*6. Highway transport*	Air
Total weight of vehicle	Vehicles	*2. Highway costs*
Distribution of weight	Weights	Construction
On wheels	Number of units	Maintenance
Sprung and unsprung	Routing	Operation
Tires	Terminals	*3. Financing highway improvement*
Speed	Financing	Methods
Wear	*7. Franchises*	Direct taxation
7. Density of traffic	Street railways	Bond issues
Width of pavement	Motor, freight, and bus lines	Vehicle taxes and licenses
	Pipelines	Tolls
	Conduits	Distribution of cost among government
	Telegraph and telephone	

significant contributions were made by some of the men present at the meeting. Among these were Dean Marston's paper entitled "A National Program for Highway Research," presented at the Annual Meeting of the American Association of State Highway Officials on December 8-11, 1919. This paper outlined the tentative plan for a national program of highway research (Table 1) under the coordinating auspices of the National Research Council, and indicated the role and the necessity of participation by the American Association of State Highway Officials, the U.S. Bureau of Public Roads, the road laboratories of engineering colleges and engineering experiment stations, municipal testing laboratories, manufacturers' research departments and associations, commercial laboratories, technical societies, and consulting engineers.

Mr. MacDonald was quite active in encouraging and supporting this cooperative research endeavor. Being conversant with all aspects of prevailing highway conditions, he was in a position to realize the urgent, growing need for a coordinated national highway research program to ensure the development of a sound and acceptable federal and state highway system.

In a discussion of Dean Marston's paper, Mr. MacDonald reviewed some of his experiences in trying to interest the executive officers of the highway departments in the program. Some of the executives were fearful that any arrangement under a national committee would cause an undue delay in getting the research results disseminated. He assured them that the results, however attained, would be made available promptly, because the Highway Research Committee of the National Research Council would be primarily a correlating body rather than an investigative agency. Dean Marston corroborated this statement.

At the suggestion of Mr. MacDonald the "Outline of the Field To Be Covered by Highway Research" was prepared in September 1920 by A. N. Johnson, College of Engineering, University of Maryland; Charles J. Tilden, Division of Engineering Mechanics, Yale University; and A. T. Goldbeck, Engineer of Tests and Investigations, Bureau of Public Roads. This first outline of a comprehensive highway research program was more of a general view of the research field than a work program.

Mr. MacDonald also presented a paper entitled "Highway Engineering Research" at the 34th Annual Convention of the Association of Land-Grant Colleges on October 19-22, 1920, at Springfield, Massachusetts. In the paper he wrote:

. . . Turning to the other major problem—that of research and investigation along highway engineering and highway transport lines—it has been recognized that a great national program is needed if the public interests are to be served by properly built and maintained highways, and if the construction of these highways is to be put upon an economically sound financial basis, so that the funds expended will prove investments returning the highest possible percentage of dividends in the service rendered. . . . For the next quarter of a century improvement of public highways will be the greatest single public activity, and will require such enormous sums of money that these expenditures should and must rest upon the soundest principles of engineering and economics.

After reviewing briefly the highway research program described previously by Dean Marston and the outline of a comprehensive program, Mr. MacDonald stated:

The National Research Council is a correlating body, and under this plan the federal bureaus concerned or those which in exercising their proper functions could lend aid to the solution of these problems, the state highway departments of the individual states, and the educational institutions could institute cooperative research activities. . . . As an example, there has just been put into effect a cooperative agreement between the University of Maryland, the Maryland State Highway Department, and the Bureau of Public Roads, by which certain problems of highway research will be taken up at once. . . .It is probable that the first of these cooperative problems will be a comprehensive study of highway traffic, and the work will be done under the immediate direction of the Dean of Engineering of the University. . . .It is my earnest hope that cooperative activities may be established in many educational institutions along these general lines to the end that a great national highway research investigational program may be instituted and carried forward without delay.

This paper produced an immediate and favorable reaction, as the Engineering Section of the Association of Land-Grant Colleges adopted a resolution that the Executive Committee be instructed to cooperate with the Secretary of Agriculture in securing funds for research in highway engineering.

These progressive activities to broaden the coordinated highway research program by Bureau of Public Roads personnel were appreciated and recognized by the Division of Engineering. Chairman Adams in his opening remarks at the November 11, 1920, organizational meeting of the Board stated: "I welcome you on behalf of the National Research Council's Engineering Division and include as hosts the Bureau of Public Roads because of its active and helpful part." Following Chairman Adam's suggestion that Dean Marston be named the conference chairman, he continued, "I also take the liberty of suggesting Mr. Curtiss of the Bureau of Public Roads as secretary. . . . Mr. Curtiss has been active in the preparation of the conference." One of Mr. Curtiss' major activities in connection with the conference was the preparation of a tentative set of bylaws for the contemplated

National Advisory Board on Highway Research. The conferees reviewed and adopted these bylaws on the afternoon of November 11, 1920.

The three representatives from the Bureau of Public Roads—Messrs. MacDonald, Curtiss, and Goldbeck—took an active part in this organizational conference. Mr. MacDonald presented one of the keynote addresses entitled: "Need for Highway Research." He also was a member of the Committee on Organization and Bylaws and contributed much to the general discussion on several matters. The following excerpt from his paper is of special significance:

> There is this point that will relate not only to the Bureau of Public Roads but to all governmental bureaus that we desire to have assist in this program. . . . Under the Executive Order of President Wilson creating the National Research Council, governmental bureaus are not only authorized but are directed to cooperate. . . . The Executive Order is very broad—it is broader probably than almost any legislation on the subject and gives great authority to governmental bureaus that they do not have under other legislation. . . .Now, unless the activity is connected with the Research Council I do not believe that authority would exist. . . .This program of highway research must be a function of the National Research Council.

Mr. Curtiss served as Secretary for the Conference, in addition to drawing up the bylaws. A. T. Goldbeck, Chairman of the Committee on Structural Design of Roads, also made a report at the Conference. Although the committee had not been definitely organized, Mr. Goldbeck briefly outlined some of the work to be conducted by the committee by reviewing the Bureau's present investigative activities in this general area.

The newly created Advisory Board on Highway Research became functionally operative primarily because of the financial support furnished by the Bureau of Public Roads. This monetary assistance is still contributed through an annual cooperative agreement between the (now) Department of Transportation and the National Academy of Sciences. The first contractual agreement was signed in August 1921, and the amount involved was $12,000 for the fiscal year ending June 30, 1922. The importance of this financial support is appreciated when it is realized that the total operating budget of the Advisory Board for the initial fiscal year was only $14,500.

In fact, during the first 24 years of the Board's existence or until the Research Correlation Service arrangement with the state highway departments was inaugurated in 1945, the financial support by the Bureau averaged about 62 percent of the operational budget. This varied from a high of 92 percent in 1928 to a low of 51 percent in 1941. The financial support from the Bureau has not only been continuous during the past 50 years, but it has greatly increased from the initial grant, as shown in Table 2.

TABLE 2 Annual Financial Support From the Bureau of Public Roads and the Total Operational Budget of the Highway Research Board

YEAR	BPR FUNDS	TOTAL BUDGET	YEAR	BPR FUNDS	TOTAL BUDGET
1922	$12,000	$14,500	1946	$ 20,000	$ 107,863*
1923	12,000	14,000	1947	20,000	130,673
1924	12,000	15,550	1948	20,000	134,722
1925	12,000	20,930	1949	20,000	155,766
1926	12,000	22,806	1950	20,000	182,099
1927	12,000	20,794	1951	30,000	187,837
1928	12,000	13,090	1952	30,000	191,457
1929	12,000	17,532	1953	30,000	193,879
1930	15,000	24,709	1954	30,000	195,000
1931	15,000	26,997	1955	30,000	221,688
1932	15,000	26,185	1956	36,000	251,800
1933	15,000	22,807	1957	36,000	253,600
1934	15,000	20,672	1958	45,000	308,519
1935	15,000	25,062	1959	45,000	329,176
1936	20,000	32,494	1960	56,250	420,030
1937	20,000	28,360	1961	61,875	454,562
1938	20,000	31,822	1962	86,875	583,393
1939	20,000	36,190	1963	86,875	759,147
1940	20,000	35,945	1964	86,875	818,543
1941	20,000	39,927	1965	156,375	900,216
1942	20,000	38,260	1966	156,375	940,040
1943	20,000	34,357	1967	156,375	1,038,894
1944	20,000	28,446	1968	322,625	1,376,000
1945	20,000	39,974	1969	322,625	1,572,000
			1970	322,625	1,675,000

NOTE: Does not include Special Projects funds nor National Cooperative Highway Research Program.

* Inauguration of Research Correlation Service. From 1945 on, most of the balance over the BPR funds was contributed by the state highway departments.

Table 2 also shows that since 1945 the increased financial support from the Bureau has kept pace in percentage with the growing financial backing received from the state highway departments for the expanding activities of the Research Correlation Service. During this interval the Bureau's contribution to the Board's operational budget has averaged about 14 percent and that from the state highway departments about 60 percent.

In addition to the financial support on the regular operational activities of the Board, the Bureau has also contributed funds and other services to many special projects the Board has administered, as summarized in Table 3. The results of practically all of these special activities are recorded in the

TABLE 3 Bureau of Public Roads Contributions to HRB-Administered Research Activities

ACTIVITY	DATE	CONTRIBUTIONS	
		AMOUNT	TYPE
Safety Research	1936-1940	$ 79,450	Funds
Road Test One–MD	1950-1952	40,000	Personnel and equipment
Intergovernmental Relations	1952-1954	24,500	Funds
WASHO Road Test	1952-1954	70,000	Supplies, equipment, and personnel
AASHO Road Test	1954-1962	6,385,100	Funds
		940,600	Supplies, equipment, and personnel
State Highway Finance and Taxation Studies	1956-1958	5,000	Funds
Highway Laws Project	1956-1961	126,000	Funds
Framework Study for Urban Research	1960	4,000	Funds
Workshop Conference on Economic Analysis	1960	1,500	Funds
Workshop Conference on Formulating Construction Program	1961	2,000	Funds
Highway and Public Utility Liaison Practices	1962	10,000	Funds
Workshop Conference on Planning in Highway Administration	1962-1964	10,000	Funds
Clearing House on Urban Transportation Research	1962-1964	17,500	Funds
Planning for Initiating National Cooperative Highway Research Program	1962-1963	12,500	Funds
Workshop Conference on Aspects of Highway Law	1963	2,500	Funds
Information Storage and Retrieval System	1965-1967	350,000	Funds
Urban Model Development	1967-1968	12,500	Funds
Conference on Transportation and Community Values	1969	20,000	Funds
Workshop on Structural Design of Asphalt Concrete Pavement Systems	1970	33,500	Funds
Conference on Movement of Goods in Urban Areas	1970	25,000	Funds

publications of the Board. Some exceptions are the original full reports under Safety Research, published by the Government Printing Office as House Document 462, Seventy-fifth Congress, third session (1938), consisting of the following parts:

Part 1. Nonuniformity of State Motor-Vehicle Traffic Laws

Part 2. Skilled Investigation at the Scene of the Accident Needed to Develop Causes

Part 3. Inadequacy of State Motor-Vehicle Accident Reporting

Part 4. Official Inspection of Vehicles

Part 5. Case Histories of Fatal Highway Accidents

Part 6. The Accident-Prone Driver

These reports were summarized in a 1938 Bureau of Public Roads publication entitled: *Highway Accidents—Their Causes and Recommendations for Their Prevention.*

Another case in which the results of an HRB-administered project were not published by the Board was "Planning for Initiating National Cooperative Highway Research Program." The project was of such a nature that no formal report was considered necessary, but subsequent reports on the project recognized the importance of this initial contribution.

The chief administrators of the Bureau (Table 4) have not only financially supported the activities of the Board continuously since it was organized, but they have also rendered valuable service as members of the Executive Committee of the Board. From February 1921, when the first Executive Committee was approved, until 1928, the chief administrator of the Bureau was elected to membership on the committee. Since that time the Bylaws provide

TABLE 4 Head of Organization Administering Federal Aid Highway Program

NAME	TITLE	TERM OF OFFICE
Thomas H. MacDonald	Chief, Bureau of Public Roads	1919-1952
Francis V. duPont	Commissioner of Public Roads	1953-1954
Charles D. Curtiss	Commissioner of Public Roads	1955-1957 [1]
John A. Volpe	Federal Highway Administrator	— [2]
Bertram D. Tallamy	Federal Highway Administrator	1957-1960
Rex M. Whitton	Federal Highway Administrator	1961-1966
Lowell K. Bridwell	Federal Highway Administrator	1967-1968
Francis C. Turner	Federal Highway Administrator	1969-

[1] Term of office as Commissioner of Public Roads overlapped appointment of Federal Highway Administrator.

[2] Interim Federal Highway Administrator, October 1956-February 1957.

TABLE 5 Sampling by Years of HRB Committee Membership From BPR

Year	Total Committees	Chairmen and Secretaries From BPR	Percent of Total	Total Committee Assignments	Members From BPR[1]	Percent of Assignments
1922	6	2	33	50	9	18
1934	36	11	30	253	40	16
1944	50	15	30	431	73	17
1954	74	19	26	989	109	11
1964	124	21	17	2,198	169	8
1968	124	15	12	2,812	153	5

[1] One individual may be a member of more than one committee; for example, the 1968 figure of 153 committee assignments from the Bureau represents 103 individuals.

that the chief administrator shall be an ex officio member of the Executive Committee.

The Bureau of Public Roads is also one of the founder organizations of the Board, and its membership has been continuous for 50 years.

In addition to the continuous financial support and services received from the chief administrators, hundreds of the technical staff of the Bureau have rendered valuable services as chairmen or members of departments and committees. Table 5 contains a periodic sample of the extent of these services and reveals the growth of the participation of the Bureau's technical staff in the Board's important committee activities, which have steadily increased to meet the challenging problems of highway technology. The data in the second and sixth columns of Table 5 show the close relationship between the growth in the number of committees and the growth in the number of committee assignments from the Bureau.

Over the years BPR technical personnel have presented hundreds of scientific papers that contributed greatly to the progress of highway technology. That these many services were appreciated by their co-workers in the highway research field is evident from the recognition that the staff members of the Bureau have received as winners of the Board's special awards, consisting of the Roy W. Crum Distinguished Service Award (originally the Highway Research Board Distinguished Service Award) and the Highway Research Board Award. To date, 17 winners of these awards have been BPR staff members. (A complete listing of the winners of awards is given in Appendix K.)

In addition to the winners of these HRB special awards, several of the Bureau's administrative staff have been the recipients of the George S. Bartlett Award, which is given to perpetuate the spirit of friendship and helpfulness to any individual who has made an outstanding contribution to highway progress. The recipient is selected by a Board of Award made up of representatives from the American Association of State Highway Officials, the American Road Builders' Association, and the Highway Research Board.

Summary

In President Wilson's May 11, 1918, Executive Order creating the National Research Council, interested governmental bureaus not only were authorized but also were directed to cooperate in the effective prosecution of the Council's work. Such a cooperative relationship between the Bureau of Public Roads, a governmental bureau, and the Highway Research Board, a unit of the National Research Council, is a prime example of the productiveness of such an arrangement. It was T. H. MacDonald, Chief of the Bureau from May 5, 1919, to April 1, 1953, who visualized the value of such a joint operation and provided the major part of the necessary initial financial support that made the Board a functionally operating agency on July 1, 1921. Mr. MacDonald and his successors have continued to give not only financial support but also other valuable assistance to the Board for the past 50 years. It is significant that Rex M. Whitton served as Chairman of the Board's Department of Maintenance and Ralph Bartelsmeyer on the Advisory Committee for the AASHO Road Test long before they assumed their major responsibilities with the Federal Highway Administration. Hundreds of other BPR staff members have also given valuable services to the Board as chairmen and members of its departments and committees and as authors of papers presented at the annual meetings. The Board is indeed fortunate to have had the aid of all of these men whose experience and standing in the highway research field command both public and professional recognition.

Undoubtedly the success of this long technical association has been due to the high regard the personnel of these two independent agencies have for each other's role in the field of highway research and the recognition that the arrangement was of great reciprocal value.

1 HRB Director Fred Burggraf took an active part in training and encouraging young engineers in the highway field. Here he chats with two students in the U.S. Bureau of Public Roads training program following one of a series of lectures given by him in 1954.

2 1958 winner of the coveted Roy W. Crum Award and long-time active worker on HRB committees, E.H. Holmes is shown (center) with K.B. Woods (left) of Purdue University. Professor Woods was Chairman of the AASHO Road Test Advisory Committee and past Chairman of the Executive Committee. On the right is Rex M. Whitton, past Federal Highway Administrator and also a past Chairman of the Executive Committee and of the Department of Maintenance. Mr. Holmes is Director of Policy Planning of the Federal Highway Administration.

3 Francis C. Turner (left), Federal Highway Administrator, U.S. Department of Transportation, and ex officio member of the Executive Committee of the Highway Research Board, has been a strong supporter of the Board throughout the years and has contributed greatly to its progress through the benefit of his experience and knowledge. Also prominent in HRB activities has been J. Burch McMorran (right), Chairman of the Executive Committee in 1966 and Superintendent of the New York State Department of Public Works. Mr. McMorran retired from public office to become a transportation consultant in Troy, New York.

1

2

3

7

Special Research Projects and Programs

As mentioned throughout Chapter 3, from the earliest days the Board was called upon to conduct special research projects. Sponsors of such research generally looked to the Board either because of some particular competence of Board staff or committees or because there was simply no other appropriate place for the research to be done. In later years, the Academy and the Board looked with disfavor on in-house research work if such work could be done by the research community at large. For this reason, the Highway Research Board has never submitted a proposal to do research work except on a sole-source basis. The Board does not compete with the research community.

Yet from the beginning sponsors have, for one or more reasons, asked HRB to conduct, either in-house or by subcontract, research work on a wide variety of subjects. The size of these projects as reflected by the amount of funds involved ranges from zero, for some done by the volunteer effort of committee members, to a maximum of over $27 million for the AASHO Road Test. Some were rather casual and candidly did not produce very much of value and others produced landmark findings and major advances in the state of highway transportation technology.

In Appendix L are listed, by date and title, those special projects with which HRB was involved up to the early Sixties. During the early Sixties, there were two major developments affecting the Board's approach to special projects. The first was the formation of the National Cooperative Highway

Research Program and the other, the formation of the Special Technical Activities Division of the Board. These programs are discussed in some detail later in this chapter. Research projects undertaken under NCHRP and under the Special Technical Activities Division are listed in Appendixes N and O respectively.

Except for a study in the early Thirties on traffic survey methods and forms, one in 1933 on evaluation of highway systems, and one on safety research sponsored by the Department of Agriculture in the late Thirties, all of the early special projects related to physical research. Highway men had much to learn about steel reinforcement, the behavior of soils, culverts, the curing of concrete, mineral aggregates, dust palliatives, concrete and asphalt pavements, and the like. From 1940 on, greater concern was shown in matters of economics, traffic control, motor vehicle size and weight, and intergovernmental relationships. However, many physical problems remained unsolved.

Foremost among the physical problems facing highway administrators, designers, and others was the relationship between performance of pavement and its structural design and traffic. During the half-century of HRB's life, no other relationship has been so elusive. Without doubt, highway administrators have spent more money on research in this than in any other single area of concern.

Libraries are filled with technical details of the problems caused by motor vehicle size and weight differentials. The problems are by no means new, as shown in Table 6, which traces principal events in the chronology of design-performance relationships.

In 1919 A. T. Goldbeck said, "At the present time it is impossible to calculate the proper thickness of a concrete road slab with much certainty." Unfortunately, 50 years and millions of dollars later his statement is still true.

In July 1919 the first of a long series of road tests was under way at Arlington Farms, across the Potomac from Washington, D.C. The Arlington tests were conducted by the Bureau of Public Roads. From 1920 to 1923 the Bates Road Test was conducted in Illinois. The Bates test involved six major types of pavement in a test road 2½ miles long and controlled wheel loads ranging from 2,500 to 13,000 lb. Depths of pavement ranged from 4 to 13 in. in 68 sections, each 200 ft long. The objective was to ascertain what type of pavement and what thickness would be required for specified loading. Two men who were later very active in Highway Research Board affairs, Fred Burggraf and A. C. Benkelman, participated in this test.

Many significant relationships were found, but the goal of a universal pavement design formula was not. The following remark from the final report on the Bates Road Test was particularly prophetic: "It is now believed

TABLE 6 Design-Performance Relations

DATE	EVENT
50 BC	Wheel load limitation of 718 lb imposed in Roman Empire
1819	Telford recommends 2,000-lb wheel load
1899	First truck manufactured
1901	Registration of motor vehicles initiated in New York (in all states by 1915)
1903	"Rules of the road" laws initiated
1904	700 trucks registered
1910	10,123 motor trucks registered
1913	First size-weight control laws instituted (Pennsylvania, Maine, Massachusetts)
1914	AASHO organized
1916	First Federal-Aid Act passed
1918	First consensus on size and weight of motor vehicles passed at joint session of AASHO and Highway Industries Association
1919	Slab impact tests conducted at Arlington, Virginia, Experimental Farm on concrete, brick, and bituminous slabs
1920	1,107,639 motor trucks registered
1920	Dynamic load test performed on circular track at Arlington Experimental Farm
1920	Highway Research Advisory Board established
1920-23	Bates Road Test conducted in Illinois
1921-22	Pittsburg (California) Road Test held
1923	Corner Formula for concrete pavement design established
1930	3,674,593 motor trucks registered
1930	Static load test of concrete strips at Arlington Experimental Farm
1932	9,000-lb wheel load for low-pressure pneumatic tires approved by AASHO and BPR
1940	4,886,262 motor trucks registered
1942	9,000-lb wheel load for low-pressure tires confirmed by AASHO
1944-54	Hybla Valley studies conducted
1945	HRB Committee on Economics of Size and Weight established
1946	Policy on Size and Weights adopted by AASHO
1949-50	HRB survey of truck weight problem in highway transportation held
1949-52	Road Test One–MD conducted at La Plata, Maryland
1950	8,598,962 motor trucks registered
1950	Loadometer-Cost-Condition studies fostered by BPR
1951-54	WASHO Road Test conducted at Malad, Idaho
1951-62	AASHO Road Test held at Ottawa, Illinois
1956	Federal-Aid Act authorizing Interstate System, "210" study, and federal limitations on size and weight passed
1960	11,914,249 motor trucks registered
1964	AASHO 1946 policy on size and weight revised
1964	House Document published on recommended sizes and weights
1968	New federal legislation on sizes and weights introduced in Congress

that truck load limitation based solely on economics of truck transportation may never fully be determined."

At about the same time, in Pittsburg, California, another test was under way. The objective of the Pittsburg test was to determine the efficiency of both reinforced and plain concrete pavements of variable thicknesses and designs on certain types of subgrade soil. The test was conducted by the Columbia Steel Company and consisted of a loop 560 ft in length containing 13 test sections. Much new information was learned here, particularly concerning the effectiveness of longitudinal joints in preventing longitudinal cracking.

The Highway Research Board did not play an important part in the tests of the early 1920's, and aside from reporting the results of many experiments throughout the 1920's and 1930's at its Annual Meetings, the Board really did not enter the picture until the advent of the Hybla Valley experiments.

Hybla Valley Nonrigid Pavement Study

The Hybla Valley Nonrigid Pavement Study was a cooperative endeavor of the Board, The Asphalt Institute, and the Bureau of Public Roads. The site of the experiment was near U.S. Route 1, about 4 miles south of Alexandria, Virginia. An oval test track was built with tangents 800 ft in length, connected at the ends with circular curves of 200-ft radius. Each tangent contained four 200-ft sections, 44 ft wide. The experiments at Hybla Valley were under the general direction of A. C. Benkelman of the Bureau of Public Roads. Frank Olmstead of the Bureau was in charge of soils studies, Ed Driscoll represented The Asphalt Institute, and W. N. Carey, Jr., the Highway Research Board. Most of the experiments involved static or very slow moving loads. Loadings through rigid plates of from 12 to 80 in. in diameter were made, and incremental and repetitive loads up to 110,000 lb gross were available. Some studies were made with flexible plates. Throughout the experimental period, careful records were maintained of soil and pavement component conditions.

That there is considerable elastic movement within a pavement structure under load was a significant finding from the Hybla Valley study. Also, a great deal was learned about instrumentation and field measurement procedures that was to be of inestimable value in later road tests.

Motor Vehicle Size-Weight Study

Directly related to the search for relationships between performance and design was a study in 1948 and 1949 of the performance of motor vehicles of various sizes and weights on different pavements, primarily as affected by grade. Field work, under the direction of Carl Saal of the Bureau of Public Roads, involved a comparison of the travel time and fuel consumption (among other things) of various truck and trailer combinations traveling between two points on the Pennsylvania Turnpike with its then-modern grades and alignments and also traveling on parallel Route 30 through a very hilly section of Pennsylvania. The highly significant results of this experiment were published by the Board in Research Report 9-A.

Road Test One–MD

The first of what was to have been a series of regional road tests, Road Test One–MD (usually known as the Maryland Road Test) was conceived by the Interregional Council on Highway Transportation in 1949. The cost of the project was borne by the cooperating agencies through contributions totaling $150,000 from the highway departments of Connecticut, Delaware, Illinois, Kentucky, Maryland, Michigan, New Jersey, Ohio, Pennsylvania, Virginia, Wisconsin, and the District of Columbia. In addition there were contributions of personnel, equipment, and services amounting to $95,000 from the Bureau of Public Roads, 7 truck manufacturers, 14 petroleum companies, and the Department of Defense. The test was conducted on a 1.1-mile section of portland cement concrete pavement that had been constructed in 1941 on US 301, approximately 9 miles south of La Plata, Maryland. The principal purpose of the test was to determine the relative effects of four different axle loads on an existing two-lane concrete pavement. Traffic tests were operated in 1950 and postmortem examination of the soils and pavement after traffic was completed in 1951. The final report, published by the Board as Special Report 4, showed that on A-6 soils, a 25 percent increase in axle load from 18,000 to 22,400 lb resulted in 7 times as much structural failure due to cracking. Similarly, 44,800-lb tandem axles caused over 12 times as much structural failure due to cracking as 32,000-lb tandem axles. Other findings relating to stresses at various points in the pavement slabs and embankment soil were reported.

The Chairman of the Road Test One–MD advisory committee was Fred Burggraf; W. N. Carey, Jr., served as Secretary. All field testing was under the direction of A. Taragin, assigned to the project by the Bureau of Public Roads. He was assisted by three engineer trainees, Theodore Dec, S. W. Smith, and J. R. Hutchins. Frank Olmstead of the Bureau was in charge of soil studies and Earl Sutherland and Harry Cashell of BPR were in charge of instrumentation and strain and stress measurements. Dozens of other engineers and administrators participated in the advisory committee, data analysis, and other aspects of this project.

The WASHO Road Test

The next in the series of regional road tests administered by the Board was sponsored by the Western Association of State Highway Officials. In 1951 WASHO requested the Board to supervise the construction and testing of a road to determine the effects of various axle loads on certain designs of flexible pavement. A site was selected in southeastern Idaho near Malad. Two test loops, each with 1,900-ft tangents, and five test sections ranging in depth from 6 to 22 in., were constructed on each tangent. The effects on these pavements of 18,000- and 22,000-lb single axle loads and 32,000- and 40,000-lb tandem axle loads were studied. Roughly a half million dollars was contributed for this test by the highway departments of Alaska, California, Colorado, Idaho, Montana, Nevada, New Mexico, Oregon, Texas, Utah, Washington, and Wyoming. The Bureau of Public Roads contributed an estimated $280,000 in the form of engineering personnel, instrumentation, photography, supplies, and equipment. Truck manufacturers loaned new vehicles for use on the test and truck trailer manufacturers loaned trailers. Fuel and lubricants were contributed by three petroleum companies. Test traffic started in 1952 and continued intermittently until May 1954. The results of the test are reported in HRB Special Report 22.

W. N. Carey, Jr., supervised the project for the Highway Research Board. A. S. Rath was in charge of construction activities. H. S. Sweet, A. C. Benkelman, and William M. Aldus each served as research engineer during part of the project. Operations engineer was Theodore Dec. Several young men who were then engineer trainees for the Bureau of Public Roads served in various capacities during the construction and field test phases. It is significant that most of these men now occupy senior positions in the Bureau of Public Roads. A partial listing includes Howard H. Boswell, Rex C. Leathers, L. E. Lybecker, W. L. Mertz, C. Robert Wright, Michael Lash, Jr., Ralph W. Gibson, Ralph T. Sigawa, Jack R. Hutchins, David M. Hall, Emery L. Shaw, and Ernest W. Harris.

Engineer observers, Richard A. Lill for the American Trucking Associations and Gordon Ray for the Portland Cement Association, worked right along with the other engineers.

The advisory committee for the WASHO Road Test was chaired by Earle V. Miller of the Idaho Department of Highways. Here again, of course, dozens of others contributed to the success of this project.

The AASHO Road Test

After the WASHO Road Test it was decided to discontinue the regional tests and to conduct instead a national test designed to include more variables than could have been studied in the Maryland or WASHO projects. This project, sponsored by the American Association of State Highway Officials and administered and directed by the Highway Research Board, was a study of the performance of pavement structures of known thickness under moving loads of known magnitude and frequency. Portland cement concrete and asphaltic pavements, as well as certain types of bridges, were included in the test facility, which was constructed especially for the test.

The project was considerably larger and more comprehensive than previous studies, and the design of the experiment contained features introduced at the request of the Academy by a team of statisticians that had not been incorporated in the other two full-scale tests. At a joint meeting of the Executive Committees of the Association and the Board in February 1955, the Board, with the approval of the National Academy of Sciences–National Research Council, agreed to undertake the administration and direction of this large-scale field research project. This $27-million project, located between Ottawa and LaSalle, Illinois (southwest of Chicago), was financed by the highway departments of the 48 states, Hawaii, the District of Columbia, and Puerto Rico. The Bureau of Public Roads, members of the Automobile Manufacturers Association, the petroleum industry, the Department of Defense, and the American Institute of Steel Construction also contributed financial support. The following table contains a brief summary of the contribution from each source:

Source	*Amount*
Joint state funds	$11,820,000
Bureau of Public Roads [1]	7,305,700
Illinois federal aid	4,042,000
Automobile Manufacturers Association	1,300,000
Department of Defense [2]	1,486,450
American Petroleum Institute	875,000
Other agencies [2]	260,070
American Institute of Steel Construction	25,000
TOTAL	$27,114,220

[1] $940,600 in contributed services.
[2] Contributed services.

The test facility included four main sections, each consisting of 6,600 ft of test sections of divided roadway, connected at each end by a turn-around of 200-ft radius, a fifth loop of the same general character with 4,000 ft of test sections, and a sixth loop of 2,000 ft of test sections with turn-arounds of 50- and 30-ft radius. Each loop had two test lanes, with concrete on one side of the median strip and asphaltic concrete on the other side. The tangent sections of the six loops contained 836 separate test sections with various combinations of surface, base, and subbase thickness. Construction of the test loops was completed in September 1958 and full-scale test traffic began on November 5, 1958, and terminated on November 30, 1960. Each test pavement surviving at the end of the test had been subjected to 1,114,000 axle load applications of a specific load. The axle loads varied over a wide range from 2,000-lb single to 48,000-lb tandem. The bridge program included 18 slab-and-beam bridges built at four locations in the two traffic loops carrying the heaviest axle loads. The selected stress levels were substantially in excess of standard design stresses.

The expenditure of $27 million represents by far the largest amount of money ever spent on a single highway research project. The project attracted worldwide attention, as evidenced by the representatives from 62 countries that were among the 14,000 who visited the test site during the construction and testing period. The final report was published in French, Japanese, and German as well as English. The project produced a reservoir of facts long needed for the development of a more refined and scientific design of pavement. Comprehensive descriptions of the project, its operation, research, and findings are recorded in eight Special Reports of the Board: 61A, *History and Description of Project;* 61B, *Materials and Construction;* 61C, *Traffic Operations and Pavement Maintenance*; 61D, *Bridge Research*; 61E, *Pavement Research*; 61F, *Special Studies*; 61G, *Summary Report*; and 73, *The AASHO Road Test*: *Proceedings of a Conference* (1962).

Throughout the construction, traffic, and report writing phase of this project, K. B. Woods of Purdue University served as Chairman of the Advisory Committee, a very hard-working group that met for two days at a time, roughly four times per year, for about 6 years, with an attendance record of over 90 percent. Literally hundreds of men served on this advisory committee, on the many special-purpose committees and panels, and in the various activities related to the research. Their names would truly be an honor roll of highway administrators and researchers. They cannot be listed here but they may be found in the Road Test reports of the Highway Research Board.

The field staff was headed by W. B. McKendrick, Project Director; W. N. Carey, Jr., was Chief Engineer for Research; W. E. Chastain, Sr., on loan from the Illinois Division of Highways, supervised construction; Peter Talovich represented the National Academy of Sciences as Business Manager. Flexible pavement research was under the direction of A. C. Benkelman, and rigid pavement research was under F. H. Scrivner. Ivan M. Viest supervised bridge research; Paul E. Irick, data processing and analysis. James F. Shook was Materials Engineer; H. H. Boswell, Maintenance Engineer; R. C.

Leathers was Engineer of Special Assignments; H. C. Huckins, Instrumentation Supervisor; and W. J. Schmidt, Chief of Public Information. Here again, it is difficult to draw the line; so many others played significant parts. Col. A. A. Wilson and Lt. Col. R. J. Lombard supervised the troops that drove the test vehicles. Many engineers were assigned by the Illinois Division of Highways and by other highway departments in the Middle West. Over 100 engineer trainees from the Bureau of Public Roads spent periods of approximately 6 months each throughout the test. Consultants and observers from other countries and from trade associations worked along with the staff personnel.

The Executive Committee of AASHO at its meeting in June 1962 took formal action commending the Board for its scientific conduct of the project. The AASHO Road Test represents an exciting chapter in the history of the Highway Research Board and one of which the Board can be proud.

National Cooperative Highway Research Program

One of the stated objectives of the Board in 1920 was "to assist in outlining a comprehensive national program for highway research." When one places this objective in context with the articulated role of the Board, it may well be restated as "to assist in preparing a comprehensive highway research *needs* outline of *nationwide* concern." Consequently, this has been a primary objective of the Board through the years, and on several occasions the Board has drafted broad outlines of needed research and encouraged organizations to undertake appropriate components of these programs. It was never considered proper for the Board itself to get into the operations except on an ad hoc basis and at the special request of the state highway departments or other appropriate organizations.

Through the years the Board has reiterated its objective for a national program and has, through its own internal mechanisms, kept abreast of the changing transportation scene and has responded to the needs apparent in the researchable problem areas. The result has been a periodic drafting of needs statements and suggested priorities (see Appendix M). In recognition of the increasing emphasis on the systems approach to transportation problems, the Board has somewhat modified its objective to permit the development of a broader program encompassing consideration of other modes of transportation and the interactions of transportation with the other aspects of society. Thus, the purpose of the Board today is to advance knowledge concerning the nature and performance of transportation systems by stimulating research and disseminating the information derived therefrom.

Perhaps the most outstanding study of needs was made by the staff and published in 1959 as Special Report 55. This pointed up both needs and benefits in highway research and showed clearly that research, which was receiving only 2 mills out of the road dollar, was not keeping pace with the problems, and that the Interstate Highway System (just getting under way) was adding new and pressing problems day by day. The impact of this report, coupled ultimately with the preparation by A. E. Johnson, Executive Director of AASHO, of a paper for the American Society of Civil Engineers on the research activities in the state highway departments—whereby it was determined that 32 states were researching the same problem—led one day to a conversation between Mr. Johnson and E. H. Holmes, currently Director, Office of Policy Planning, Federal Highway Administration, U. S. Department of Transportation, wherein the concept of the National Cooperative Highway Research Program (NCHRP) was born. It was thought that the Highway Research Board, because of the adequacy demonstrated in managing the AASHO Road Test and because of its recognized objectivity and understanding of modern practices, would be the logical agency to administer such an endeavor.

Furthermore, it was considered important that the Highway Research Board could call on the tremendous reservoir of professional competence represented in the divisions of the National Research Council and throughout the National Academy of Sciences for advice and counsel and expertise in almost any subject with which it was likely to become involved in such a program.

Nevertheless, this was to be an operational research program and as such was viewed with considerable misgivings by senior members of the Governing Board of the National Research Council. It was only after eloquent expression by Augustus Kinzel, then Chairman of the Division of Engineering and Industrial Research, of the need for the research program and of the lack of another suitable home for the program that the Governing Board agreed to permit HRB to undertake it.

On June 19, 1962, the concept was brought to fruition when a three-way agreement was signed by the National Academy of Sciences–National Research Council, the American Association of State Highway Officials, and the Bureau of Public Roads to inaugurate a program of systematic, well-designed research to provide the most effective approach to the solution of many problems facing highway administrators and engineers. The first research program was received at the same time and consisted of 34 problems. Each year since, AASHO has proposed specific problems for inclusion in the Program's fiscal year activities. At least two-thirds of the participating member departments

HIGHWAY RESEARCH BOARD

NATIONAL COOPERATIVE HIGHWAY RESEARCH PROGRAM

Committee Structure

HRB EXECUTIVE COMMITTEE

ADVISORY COMMITTEE

RESEARCH FIELD A: ADMINISTRATION

PROBLEM AREAS:
- 2 - Economics
- 11 - Law
- 19 - Finance

RESEARCH FIELD B: TRANSPORTATION PLANNING

PROBLEM AREAS:
- 7 - Traffic Planning
- 8 - Urban Transportation

RESEARCH FIELD C: DESIGN

PROBLEM AREAS:
- 1 - Pavements
- 12 - Bridges
- 15 - General Design
- 16 - Roadside Development
- 22 - Vehicle Barrier Systems

RESEARCH FIELD D: MATERIALS AND CONSTRUCTION

PROBLEM AREAS:
- 4 - General Materials
- 9 - Bituminous Materials
- 10 - Specs., Procedures, and Practices
- 18 - Concrete Materials

RESEARCH FIELD E: SOILS AND GEOLOGY

PROBLEM AREAS:
- 21 - Testing and Instrumentation

RESEARCH FIELD F: MAINTENANCE

PROBLEM AREAS:
- 6 - Snow and Ice Control
- 13 - Equipment
- 14 - Maintenance of Way and Structures

RESEARCH FIELD G: TRAFFIC

PROBLEM AREAS:
- 3 - Operations and Control
- 5 - Illumination and Visibility
- 17 - Safety

SPECIAL PROJECTS

PROBLEM AREAS:
- Hwy. Research Information Service
- Research Needs in Highway Transportation
- Traffic Capacity and Safety
- Future Transportation Preference
- Highway Information Synthesis
- Right-of-Way and Legal Problems
- Special AASHO P & DP Comm. Research
- Hwy. Design Graphic Systems

must approve the research problems and agree to their financial support before they can be brought into the Program. Following a balloting by the member departments, the approved problems are referred to the Academy where the chairmen of NCHRP advisory panels and members of the NCHRP Advisory Committee review each yearly program to determine its acceptability for administration by the Academy through the Board. Certain projects bear the request that the Board conduct the research work with its own staff, and five such projects have been referred to date. Each state annually contracts with the Academy to commit 4½ percent of its 1½ percent federal-aid highway planning research (HPR) funds. From these contributions, a cooperative pool of about $3½ million is made available for NCHRP's contract research and for its administrative and technical operation.

Once accepted, the problems making up the program are assigned to advisory panels that are made up of persons knowledgeable in each particular problem area and who advise on technical aspects of the problem. There are presently some 350 members on these panels coming from 41 states, the District of Columbia, and Canada. They analyze the problems, outline particular projects and their objectives, and then prepare research project statements on which proposals are solicited from qualified private and public research agencies. They review the proposals, recommend contract awards, and provide counsel to the NCHRP staff responsible for surveillance of work under the research contracts. Finally, they review final reports for acceptability and for accomplishment of the approved research plan. The accompanying chart shows the structure of NCHRP.

A professional staff is assigned to NCHRP by the Board. The first Program Director was M. Earl Campbell. He was succeeded on October 15, 1964, by William A. Goodwin. In turn, Mr. Goodwin was succeeded on September 1, 1967, by the present Program Director, Krieger W. Henderson, Jr. Projects engineers with individual specialties and training in the broad areas of physical research and traffic planning are responsible for administrative and technical surveillance of the contracts. In addition to reviewing quarterly progress reports and monthly progress schedules and maintaining telephone contacts, each engineer visits his assigned projects throughout their contract periods. He discusses with each principal investigator the project's status to learn if the research is being pursued in line with the approved research plan. Meetings involving the staff, advisory panel, and agency personnel are held frequently for the purpose of reviewing project progress and providing guidance for continuing work; 39 such meetings were held in fiscal year 1970. Finally, the projects engineer and the advisory panel evaluate the completed research to determine the degree of technical compliance with the contract and the acceptability of the final report to the Board and Academy.

The research findings are published in a special NCHRP report series. Each highway administrator receives a copy immediately on publication, and some 3,000 copies are formally distributed through the Highway Research Board's selective distribution system.

Still another means for bringing research findings before the practicing engineer was instituted in 1968 as the *NCHRP Research Results Digest*—a series of flyers published at frequent intervals in the interest of providing an early awareness of the research results emanating from the various projects. By making these results known as they are developed and prior to publication of the final reports, it is hoped that their early use in practice will be encouraged.

In 1970 NCHRP received its ninth program (FY 1971), consisting of 21 projects totaling about $2.8 million. From all programs 173 research projects have thus far resulted, with a total funding obligation of about $22 million (see Appendix N). The contracts have been let to agencies headquartered in more than 26 states and the District of Columbia. Universities have received about 41 percent of the contracts, nonprofit institutions 24 percent, consulting and commercial firms 30 percent, and industry, individuals, and state highway departments 5 percent.

A total of 94 interim or final research reports have been published documenting the research findings of the contracting agencies. In 1969 NCHRP found it necessary to begin a new series of publications, *Synthesis of Highway Practice*, to consist of state-of-the-art information rather than research results. Quarterly progress reports are furnished each year as the primary means of keeping the sponsors informed of all administrative and technical progress. In addition, the Program in 1966 began publishing an annual summary of progress for each year, and this document also receives very wide distribution through the selective distribution system of the Board.

The NCHRP is a unique contract research effort designed to respond quickly and efficiently to the needs of the state highway departments through the solution of the pressing operational problems in the realm of highway transportation. Although the Highway Research Board administers the Program, the research content is solely the prerogative of the American Association of State Highway Officials and the participating member state highway departments. The Program is one of applied rather than basic research, and every possible effort is made to help administrators put the findings to early use. Although research in the NCHRP is presently sponsored solely by the member departments of AASHO, the Program is designed to administer research for other agencies as well.

Special Projects Division

In recent years most research projects that are specially funded from outside sponsors or from NCHRP and conducted by staff employed by the Highway Research Board have been assigned to the Special Projects Division.

The Special Projects Division, headed by Paul E. Irick, administers and sometimes conducts a relatively small number of projects that are funded outside the regular operating budget of the Board. In the Board's new organization this division receives guidance from a ten-member Division B Steering Committee, chaired by Harmer E. Davis.

There are essentially three routes by which a given HRB activity may be designated as a Special Project:

By agreement between HRB and another National Academy of Sciences–National Academy of Engineering unit that has been asked to carry out the special study. An example is the project on Ambulance Design Criteria that was conducted in cooperation with the NRC Division of Medical Sciences.

Through HRB cooperation with sponsors, including NCHRP, of projects that require unique HRB capabilities, but that are not appropriate to the Technical Activities Division. Examples are the Highway Litter Study and the NCHRP project on Economic Consequences of Highway Improvements.

Through HRB solicitation of sponsorship, including NCHRP, for a project whose output will be directly applicable to the Board's overall objectives. The TRIS I project and NCHRP project 20-5 are examples.

In accordance with National Academy of Sciences–National Academy of Engineering policy, an advisory committee of appropriate experts may be formed to give advice and counsel on the conduct of a special project. Staff as needed are employed for each project and other designated members of the HRB staff may serve as technical advisors.

Fourteen special projects that have been undertaken since 1963 are summarized in Appendix O. The first seven were completed prior to fiscal year 1969 or soon thereafter, the eighth was completed in early 1970, and the ninth was to be completed in late 1970. The remaining five projects represent continuing work.

Four of the completed projects have dealt with guidelines and criteria for further research and development; they were concerned with pavement performance, bus design, driver education, and ambulance design. Four projects have been or are concerned with the synthesis and publication of information in particular areas of highway research, such as street utilization, economic consequences of highway improvements, current highway problems, and right-of-way and legal problems. One project included data acquisition and analysis on the subject of highway litter composition.

The remaining five projects all deal with the development of transportation research information services. The first of these was the Highway Research Information Service (HRIS), which became operational in 1967.

By capitalizing on the HRIS experience and because of the Board's concern for transportation research information beyond the highway field, new capabilities are being developed for the storage, retrieval, and dissemination of maritime information, transportation noise information, highway safety information, and transportation research-in-progress information. The last of these projects was a pilot study that was completed in early 1970, but a proposal for its continuation has been submitted to the Office of the Secretary of Transportation. Preliminary discussions have been held with the Federal Railroad Administration and the Association of American Railroads on the possibility of a Railroad Research Information Service.

Right-of-Way and Legal Problems

The NCHRP project on right-of-way and legal problems is staffed by the HRB Counsel for Legal Research and his assistants to conduct legal research and prepare papers on highway problems that involve right-of-way acquisition and control. Individual state experiences are compared and made available for possible assistance to other states.

Papers have been published on five topics, as follows:

1. Standing to sue for purposes of securing judicial review of exercise of administrative discretion in route location of federal-aid highways.
2. Relocation assistance under Chapter Five of the 1968 Federal-Aid Highway Act.
3. Valuation changes resulting from influence of public improvements.
4. Advance acquisition under the 1968 Federal-Aid Highway Act.
5. Valuation in eminent domain as affected by zoning.

Current research is being conducted on two papers relating respectively to joint development and multiple use of highways and environmental factors as affecting highways.

Synthesis of Existing Information on Highway Problems

The objective of the NCHRP project on synthesis of existing information on highway problems is to bring together research results, published material, and current practices on relatively narrow highway topics. Technical panels and consultants assist the project staff in acquiring, evaluating, and organizing the material on each topic into booklet form. Reports on four topics have been published as Syntheses on Highway Practice: *Traffic Control for Freeway Maintenance, Bridge Approach Design and Construction Practices, Traffic-Safe and Hydraulically Efficient Drainage Practice, and Concrete Bridge Deck Durability.*

Material is being developed for six additional topics. Two of these are nearly ready for publication, on scour at bridge waterways and on project scheduling and monitoring. Other topics deal with skidding, pavement rehabilitation, embankment construction, and maintenance personnel.

Maritime Research Information Service (MRIS)

The MRIS project is under the guidance of an advisory committee of the National Research Council, the Maritime Information Committee. Acquisition, selection, and dissemination activities are carried out by the MIC staff. Highway Research Board staff and information system capabilities are used to carry out storage and retrieval operations. The developmental issue of the MRIS Bulletin was published in June 1970.

Transportation Noise Research Information Service (TNRIS)

The TNRIS project has three objectives: (a) to develop a storage and retrieval service for information on transportation noise with primary emphasis on ground transportation, (b) to prepare synthesis reports on special topics within the TNRIS scope, and (c) to advise on areas of needed research and development. A Highway Research Board Advisory Committee has been formed to assist in attaining these objectives. Initial output was scheduled for late 1970.

Highway Safety Information Service (HSIS)

In the HSIS project the Highway Research Board provides the National Highway Safety Bureau with storage and retrieval service for information that is acquired and selected by NHSB, including bibliographic citations that have been published in *Highway Safety Literature*.

Special Conferences and Workshops

It has been found that special conferences and workshops, designed to explore relatively narrow specific subject areas, are highly effective as a means for advancing the state of knowledge in the subject area and disseminating that knowledge. So, in addition to its Annual Meetings, held each January in Washington, and its summer meetings, which have been held in western states, the Highway Research Board has sponsored, cosponsored, and participated in a number of special conferences and workshops in the decade of the Sixties.

Some of the workshops have been conducted within the Special Projects Division of the Board, but most of them have been conceived by committees of the Board in the regular Technical Activities Division. Some are sponsored by government or other interested agencies and some are financed completely by registration fees paid by the delegates to the conference, but all are financially independent of the other activities of the Highway Research Board.

Although there is variation from conference to conference, a rather typical workshop format involves 2½ days. The first half day consists of a plenary session in which major position papers outline the state of the art and the principal questions facing the conference. The next day and a half is occupied by breaking the group into small panels (approximately 15 persons each). Each panel addresses one facet of the general problem and prepares a report. On the final half day, in another plenary session, the panel chairmen offer their reports for the consideration of the entire group. Of course, the ultimate success of such conferences depends on the advance work of the planning committee, panel chairmen, and major speakers.

The workshops in which the Highway Research Board has been involved are listed in Appendix P.

1 The National Advisory Committee of the AASHO Road Test comprised an honor roll of prominent individuals from highway research of the 1950's. The Committee was chaired by K.B. Woods (extreme left, back row), Head of the School of Civil Engineering, Purdue University. Also in the group are R.E. Livingston, P.E. Irick, A.C. Benkelman, Fred Burggraf, J.B. Hulse, E.A. Finney, W.C. Williams, R. Bartelsmeyer, C.E. Fritts, Rex Whitton, F.N. Hveem, W.C. Hopkins, R.E. Fadum, M.S. Kersten, J.M. Griffith, H.O. Thompson, E.H. Holmes, A.E. Johnson, L.K. Murphy, J.O. Izatt, W.F. Abercrombie, W.E. Chastain, Sr., R.L. Peyton, A.A. Anderson, T.E. Shelburne, R.E. Jorgensen, H.A. Flanakin, L.C. Lundstrom, T.F. Creedon, D.K. Chacey, H.F. Clemmer, S. Goldin, C.F. Kossack, J.F. Shook, W.N. Carey, Jr., C.F. Rogers, G.D. Campbell, R.A. Lill, A.S. Wellborn, F.N. Finn, R.A. Moyer, W.B. McKendrick, Jr., R.C. Leathers, H.H. Boswell, G. Egan, W.J. Schmidt, B.E. Colley, E.J. Felt, and B.W. Marsh.

2 Experts from all areas of transportation research participated in the "Access to Airports" Engineering Foundation Research Conference near Milwaukee in the summer of 1968. HRB was a cosponsor. Pictured are, left to right, front row: S.S. Cole, T.B. Davinroy, C.R. Julian, S.G. Lardiere, K.E. Cook, L.F. Bergmann, G.E. Katzmar, R.F. Baker, and D.D.Mikesell. Second row: T.K. Jordan, R.E. Dudley, M.N. Danielson, E.N. Hall, D. Brand, R.S. Foote, Z.A. Nemeth, W.L. Garrison, and M. Herrin. Third row: E. Allen, P.W. Dygert, W.N. Carey, Jr., S. Russell, D.C. McGrath, L. Stafford, and D.M. Clancy. Fourth row: W.B. Williams, F. Koomanoff, G.P. Fisher, K.W. Shiatte, W.H.T. Holden, A.E. Brant, Jr., and C. Pearlman. Fifth row: R.A. Walbrecker, J.A. Scott, W. Rainville, F.E. Jarema, A.M. Voorhees, J.E. Snell, M.L. Manheim, E.F. Cleary, and A.J. Gellman.

1

2

1 One of the earliest full-scale test projects carried out in the United States was the Bates Road Test, which was conducted in Illinois from 1920 until 1923. Fred Burggraf and A.C. Benkelman, who were both to become prominent in HRB activities, participated in the project.

2 Test loops for the WASHO Road Test, conducted from 1952 until 1954, were specially built near Malad, Idaho. Purpose of the test was to determine the effects of various axle loads on certain designs of flexible pavement.

3 Road Test One–MD (also known as the Maryland Road Test) was conducted in 1950. Among the observers pictured here are H.S. Fairbank (standing, third from left), then Director of Research, U.S. Bureau of Public Roads, and A. Taragin (kneeling, center), who was loaned to

1

2

3

the Highway Research Board as project director by the U.S. Bureau of Public Roads. The test was conducted on a nine-year-old section of pavement near La Plata, Maryland.

4 The magnitude of the $27 million AASHO Road Test is apparent in this aerial view of half of the test layout near Ottawa, Illinois. Some of the trucks used in the test can be seen parked on the inner side of the loop in the foreground.

5 The WASHO Road Test, which ran from 1952 until 1954, was conducted for the Highway Research Board by W.N. Carey, Jr. (left), who later became Director of HRB. A.C. Benkelman (center) was research engineer. Fred Burggraf (right) was Director of HRB when he visited the project site in Malad, Idaho.

6 The Hybla Valley Nonrigid Pavement Study was carried out in 1949 near Alexandria, Virginia. Ed Driscoll, representing The Asphalt Institute (sponsors of the test in cooperation with the Highway Research Board and the U.S. Bureau of Public Roads) is shown (center) supervising a plate bearing test.

7 Frederick Seitz as President of the National Academy of Sciences was a prime mover behind the formation of the National Cooperative Highway Research Program, and his encouragement of this concept made it possible for the Highway Research Board to undertake administration of the program in 1962.

4

5

6

7

8

Some Breakthroughs in Highway Research

After an institution has operated for a half-century, one can justifiably ask: What are the results? And in the case of the Highway Research Board, which has been right in the center of highway research activities in the United States since 1920, one has the right to ask: What are the significant breakthroughs in highway research? What are the important developments in the highway field? What revolutionary or impressive discoveries can highway research point to?

Roy W. Crum had this to say:

> The developments of highway technology, although tremendous in scope, have taken place almost imperceptibly over many years. It is difficult at any one instant to point to a revolutionary discovery. Our aim must be not to produce the highway of the future overnight, but to accelerate the type of work which has brought us from the horse-and-buggy highways to the great American highway system of today.

As expressed in an editorial of the *Engineering News-Record*:

> Research knowledge in highway engineering grows by accretion. In the unfolding years of activity of the Highway Research Board, no single discovery has claimed leading notice. But each year has seen a thin layer added to the pearl of assured

knowledge placed at the service of highway transportation through the activities of the large number of laboratory and field investigators who make up the Board.

Nevertheless, there have been some revolutionary concepts and discoveries in highway technology, perhaps not sensational, yet quite substantive and significant. Among them are the following:

Bituminous and portland cement concrete pavements
Soil stabilization
Smooth, skid-resistant surfaces
Air entrainment
Controlled-access highways
Divided highways and one-way systems
Network signalization
Highway capacity determinations
Land-use traffic generation and traffic assignment
Socioeconomic analysis in highway location
Dynamics of traffic flow
Energy-systems concept in research

It should be stated that the Board does not lay claim to all of these concepts and discoveries. In some cases the Board made the discoveries, but in other cases the Board mediated the event or was simply aware of the evolution. It is not always possible to delineate clearly the boundary that marks the Board's involvement in this evolution. In most of the cases cited the author of the breakthrough conferred with the Board (its staff and committees) about his work, encountered interest and encouragement, and reported his findings through the Board's publications.

The staff of the Highway Research Board in 1966 reviewed the accomplishments made in the highway research field during a period of 50 years. The major findings were incorporated in a paper ("The Anatomy of Highway Progress") presented by D. Grant Mickle, then Director of the Board, at the Third Biennial Conference of the Australian Road Research Board in Sydney, in September 1966. The breakthroughs pointed out in that paper, as well as some other developments, are mentioned here in annotated form. Fuller details can be obtained from other sources. The advances are treated in terms of major subject areas.

Design

1. Controlled Access—A 15-mile section of the semi-experimental Bronx River Parkway in New York was completed in 1925.

2. Divided Highways—The Du Pont Parkway in Delaware pioneered this development in 1915.

3. Aerial Photography and Photogrammetry—This development accelerated in the Thirties. It is now used in nearly every kind of field survey—route location, property surveys, traffic studies, and many others.

4. Prestressed Concrete—The concept came from Europe; the first practical application in the United States was in the Walnut Lane Bridge in Philadelphia in 1949.

5. Erosion Control—One of the early (1935) committees of the Highway Research Board (joint with AASHO) worked on this problem. Agricultural science has developed techniques in various kinds of plantings to control highway erosion and provide beauty to the roadside.

6. Electronic Surveying Instruments—During the past decade, several varieties have been produced that speed up the measuring process and provide more accuracy in long-distance measurements.

7. Geometrics and Surfaces for Safety—A better understanding has been evolving since the Twenties of the mechanics of vehicular motion as related to curves, grades, and surface conditions. This has led to alerting devices as well as methods to prevent traffic accidents and to ameliorate their consequences.

8. Structural Design of Pavements as Layered Systems—The several road tests administered by HRB and others, plus satellite tests in several states, together with laboratory studies, have provided a wealth of data for assisting in structural design and prediction of service life of a pavement.

Materials and Construction

1. Air-Entrained Concrete—Discovered accidentally (circa 1945), this development has provided concrete of more durability through cycles of freezing and thawing and chloride action.

2. Prevention of Alkali-Aggregate Reaction—Detection of the active cements and aggregates and the development of inhibitors have been of remarkable benefit. Early research began about 1940 in California.

3. Slip-Form Paving—The speeding up of the paving process and the reduction in costs, together with a smooth-riding surface, have evolved during the past decade or so.

4. Automated Asphalt Pavers—This development permits the paving machine to follow automatically the predetermined grade lines by means of computer programs. This development took place during the Sixties.

5. Nondestructive Testing of Material—Testing in situ and withdrawn

specimens through sound waves, nuclear energy devices, and other devices permits the material tested to continue as a component part of a structure or to be retained as a control specimen in further research.

6. New Pavement Design Criteria—The effects of various magnitudes and frequencies of loads on concrete and bituminous pavements of various thicknesses and with controlled parameters were determined in three road tests. These relationships permit more precise design of pavement structures.

7. Pavement Performance or Service Life—The development by HRB on the AASHO Road Test of the pavement serviceability index provides a guide for determining the remaining life in a pavement and for predicting when resurfacing will be needed.

8. Quality Control in Construction Practices—A more uniform product is afforded by making use of statistical methods in the selection of samples in design and inspection.

9. Job Management—The analysis of requirements, men, materials, and machines on a time-space continuum for efficiency and effectiveness has resulted in high-capacity production.

10. Performance Testing for Acceptance of Materials—Traffic paints, as an example, are applied and tested under traffic for durability and effectiveness.

11. Beneficiation of Marginal Aggregates—With many good aggregates in short supply this practice provides for acceptable substitutes.

Maintenance Practices

1. Thin Overlays—Wearing surfaces of ½-inch thickness or less, plant-mixed and machine-laid, have been developed within the past decade. These have proved effective in providing smooth riding surfaces, reducing spot failures, and providing an economical remedy for pavements that are slippery when wet.

2. Quick Repair Procedures—Equipment and new developments in materials permit rapid repairs on heavily traveled highways. For example, concrete patching can be made and traffic restored in 3 hours.

3. Snow and Ice Control—In 1920 this was not recognized as a problem of great consequence, but today users are demanding bare pavements even as it continues to snow. With advances in snow and ice control methods, materials, and machines, bare pavements are being provided on a large mileage of key highways and streets.

4. Protection of Concrete Pavements and Bridge Decks Against Chlorides—Besides air-entrainment, the use of a 50-50 mixture of boiled linseed

oil with mineral spirits gives decided protection against deterioration from chlorides used in snow and ice control.

5. Equipment for Special Maintenance Tasks—There have been numerous advances in equipment research for many specific types of repetitive operations.

Soils and Foundations

1. Soil Stabilization—For bases and surface courses much progress has been made, particularly in the use of lime, cement, bitumen, mechanical means, and various other means in stabilization of soils. Compaction as a form of stabilization has made remarkable strides.

2. Synthetic Aggregates From Shales, Clays, and Gumbos—Lightweight aggregates for surfacing and for bridge decks have been produced through the firing of clays, shales, mine tailings, and gumbos. These have produced economies in construction costs. This development—not new in concept—has gained momentum in the past score of years.

3. Density-Moisture Concepts and Control—Relationships among moisture content, compactive effort, and soil density have been determined, thus affording significant help in design and construction practices. This development dates from the late Twenties.

4. Embankment Stability—The effect of pore water pressure on loaded cohesive soil masses and its relationship to stress-strain relations (in particular, shear) led to techniques for quick consolidation procedures in embankment construction. This research has been in progress since 1935.

5. Frost Action—Research on frost action in soils led to the prediction of performance of various soils affected by frost action, and also led to the development of devices for measuring the load-bearing capacity of soils during the thawing period. This work has been in progress for more than a score of years.

6. Supersonic Power Uses—In penetration or cutting of soil and in mixing operations, this new technique is showing great promise.

Finance and Administration

1. Incremental Method of Tax Allocation—The development of this method for highway use during the past 20 years provides a method of allocating taxes according to the costs occasioned by the different user groups, and provides a basis for a more equitable distribution of taxation.

2. Motor-Vehicle Operating Costs—Though not entirely completed, much work has been done on the cost of operating different weights of vehicles on grades, curves, and level tangents at different speeds. This provides a feedback for design. Such studies began in 1920 and are continuing.

3. Highway Organization and Management Studies—The principles of administration (policy, execution, functions, and structure) have been continuously studied and reported in HRB publications. The job management function has also been examined and reported.

4. Adequacy Ratings—The measurement of the condition of a road in terms of safety, structure, and service was given a decided impetus following World War II in the inauguration of a method known as "sufficiency ratings." Virtually all state highway departments use some form of this method as a guide to programming construction.

5. Mathematical Models—The modeling of problems for systems analysis and solution by electronic computer has advanced. Because value scales are often inadequate, models are readied for scalar measurements. This development followed the development of the electronic computer in 1946.

6. Administrative Planning—What is done with the mass of information once it is assembled so that it will lead to wise choices of action? This question is the basis for administrative planning. Many techniques are now available to assess alternative choices in terms of both social and economic consequences. The newer techniques of cost-effectiveness, CPM and PERT, PPBS, programmed ticklers, planned communication and coordination, and planned application of these useful new methods have helped greatly in choosing and timing the wise course of action.

7. Highway Administration—The concept of highway administration as two coordinate components, policy (objectives) and execution (technological operations), is gaining in recognition. The distinctive roles of each and their complementary functions need more study.

Traffic Operations

1. Highway Capacity Studies—These studies began in desultory fashion when the Board was created, then in an organized way in 1930. The first *Highway Capacity Manual* was published in 1950. Translated into nine other languages, it has been distributed throughout the world. A completely revised edition was published in 1965.

2. Traffic Control Signals—The manually operated semaphore, first used about 1910, gave way to the three-color electric signal inaugurated in New York City in 1918. Computerized network systems automatically responsive to traffic flow are now in operation mediating the right-of-way.

3. Reflective Signs, Delineators, and Markings—These date from the mid-Thirties and have been a boon to safer night driving. They are still evolving, and there is hope now for reflective markings that are clearly visible in rainy weather.

4. Ramp Metering at Interchanges—To sustain a desired level of service, on-ramps are metered to provide the appropriate number of entering vehicles on expressways.

Urban Transportation Planning

1. Transportation and Community Values—With 600 miles of the Interstate System yet to be constructed in urban areas, transportation planners anticipate some harmful socioeconomic impacts and are attempting, frequently with the help of the affected community, to use transportation facilities as instruments of creative change.

2. The Multidisciplinary Approach—Comprehensive, coordinated studies in Chicago, Philadelphia, Pittsburgh, New York, and Los Angeles, as well as other metropolitan areas, have proved the effectiveness of using the multidisciplinary approach.

Legal Studies

1. Use of Air Rights—Development of titles to air rights and tunnel rights was necessitated by increased demand for land above highway facilities and below existing structures.

2. Valuation Techniques in Appraisal of Leasehold Estates—Three techniques were developed for valuating leasehold estates: market approach, in which similar leasehold estates were in the market; cost approach, based on physical improvements made to the estate; and income approach, in which the leasehold estate is compared to the income stream.

Conclusions

Many more so-called breakthroughs could have been added, and those noted could have been rearranged under the following categories:

Safety (signing, designing, maintaining, protecting)
Economy (operation, durability, obsolescence)

Time savings (freedom from frictions of flow)
Convenience (location, access)
Comfort (including freedom from apprehension)
Aesthetics (geometrics, beautification)
Amenities (preservation of values to user and abutter)

One of the most important evolutions has been that of the research method itself. With an array of gadgets for each form of energy; with proven techniques for the conduct of controlled research; with the expanding knowledge in mathematical statistics for use in sampling and modeling and analysis; with electronic computers for the reduction of great masses of numbers; in short, with facts, techniques, and accelerated mechanics, a revolution has been produced in the research field. The Highway Research Board, with its rigorous disciplines, has augmented this development. The Board itself is a breakthrough in the highway research endeavor in the United States.

The breakthroughs in knowledge have been impressive. The greatest breakthrough still awaits discovery: the breakthrough in application. This breakthrough is also a cooperative endeavor. The researcher must interpret his findings for use in real-life situations. The technologist must recognize the benefits to be derived from changing traditional practice to something different. The administrator must authorize the change. There are many markers along the advance in knowledge in highway administration and technology. It is time that these markers led to the milepost of full usage.

DIRECTORS OF THE HIGHWAY RESEARCH BOARD—PAST AND PRESENT

Charles M. Upham
1924-27

D. Grant Mickle
1964-66

W. K. Hatt
1922-23

Roy W. Crum
1928-1950

W. N. Carey, Jr.
1967-

Fred Burggraf
1951-63

9

Motif for the Future

A history is intended to document and, where appropriate, to honor the past. Yet the history of an organization cannot be said to finish with a 50th anniversary, so it seems appropriate to look to the future.

Surely it is clear from examination of this history of the Highway Research Board and from a look around in today's world that the problems of transportation, some of which existed 50 years ago, others of which developed with changing distribution of population and changing goals of society, have not been solved. Perhaps these problems are not subject to solution, but we have learned much about them and, hopefully, we can progress far enough at least to live with them.

But certainly there is still need for research and there is still need for a central agency that can identify research needs and assign priorities, that can correlate and assist in the research effort, and that can serve as a medium for exchange of information during the time of the research program and after the results have become available. The Highway Research Board, of course, is such an agency.

In recent years, the sponsors of the Highway Research Board have made it quite clear that they expect the Board to continue its efforts of the past few years to broaden its base of subject interest. They have stated through their representatives on the Executive Committee that present HRB funds may be used for committee activity, for meetings and seminars, and for staff in transportation subject areas beyond highway transportation. They have cautioned that highway funds are restricted to situations where highways interface or where they interact with other modes of transportation or where they interact with the social and economic well-being of the community that transportation is designed to serve. It now seems likely that funds will be forthcoming from air, rail, pipeline, and other mode sponsors on a continuing and substantial basis. When this develops, the Board can truly become transportation-wide in its attempts to serve as a central clearinghouse and forum for transportation research.

Granted, then, that in 1970 there exists a need for a Highway Research Board or a broader transportation board in the same model. Let us look at some of the issues that must be faced in the next decade or two.

In order to put the needs of the future into rational focus, the staff of the Highway Research Board has listed ten principal areas of current and potential concern that should be receiving significant attention. It seems reasonably sure that in this list will be found the basis for emphasis in committee and program activity for the Board, at least for the near future. The ten areas are not listed in order of priority since the problems and their elements interact heavily with each other.

The effective use of highway funds and resources—Because the HRB staff believes that highway needs will continue to be greater than the ability of the states and federal government to generate revenues, there will continue to be a vital obligation to use highway funds and resources most effectively. This concept is retroactive in the sense that something should be done to take better advantage of the millions already spent for research.

Improving the quality of the environment through highway design and location—Because there is growing concern that modern technology, including highways and automobiles, is adversely affecting the quality of life and disturbing ecological balances, it is probable that the quality of the environment will receive substantial attention, especially in urban areas.

Integrating the different transportation modes into a unified demand-sensitive system—The increasing trend toward state departments of transportation indicates growing interest in the integration of the different transportation modes into a unified national, state, and local transportation system.

Increasing the safety of highways for users and for the community—Concern for highway safety is stabilizing after a strong initial thrust. We may expect to see a continuing steady emphasis aimed at making both highways and vehicles safer. Highways have come a long way in this respect, particularly interstate highways. There is still a great need for improvement in the primary and secondary systems. Much emphasis will be placed on the vehicle as a direct element of the system and as a mechanism for providing the driver with safeguards that he needs in the event of an accident.

Increasing capacity and reducing congestion on existing highway facilities—Public resistance to highways in some urban areas emphasizes the need for means to relieve urban traffic congestion other than the construction of new highway systems. Although it seems unlikely that there will be any major new developments in the theory of traffic flow that will dramatically increase the capacity of existing highway systems, there is a great need to apply existing knowledge and technology to urban street systems.

Improving durability of pavements and structures and reducing maintenance costs—Improvements in construction, materials, and design to increase the life and performance of pavements and structures are needed in order to protect the large investments in highway facilities. Maintenance costs must be reduced through improved management, equipment, and techniques.

Evaluation of the economic and social influences on highways, including demand, costs, and pricing—Since the end of World War II, response to demand was the primary criterion for determining highway programs. Needs were established through a combination of traffic demands, design criteria, and safety improvements. Now, economic and social influences of highways will receive greater consideration.

While social considerations may impede the construction of new highways and freeways, we do not believe that there will be a substantial shift in the use of transportation systems in urban areas to alter highway needs appreciably. Whether the vehicle be bus or automobile, fast, efficient, and safe highway corridors will still be necessary. Use of congestion-pricing techniques would only shift traffic that exceeds the capacity of the system to other points of the urban area, leaving the existing routes saturated to capacity. Because of investment and operating costs and the need for high density along the routes, mass rail transit systems will not offer economically or socially feasible solutions to urban traffic congestion, except in a few large, high-density cities. However, the integration of highways into a total transportation planning process will mean that all systems will be more closely evaluated as to responsiveness to demand in comparison to cost for the facilities.

Defining governmental, industrial, and consumer responsibility for transportation—One of the primary problems in transportation is defining the governmental relationship and responsibilities for financing, constructing, and operating the different systems. If any sufficient innovations in transportation are to be achieved, the multiplicity of governmental and other agencies responsible for planning and transportation in urban areas will have to be effectively coordinated. Councils of governments, such as are being formed in many urban areas, provide one means for synthesizing the many affected agencies into a single agency. Increased attention will be given in coming decades to the sources of funds for transportation systems and the effects that fund structure has on influencing the development of the transportation system.

Development of new transportation systems and technology—The urban environment is undergoing rapid social and economic change, and it is difficult at the present to predict the kind of transportation systems needed in future decades. It would be naive to expect that new technological hardware by itself could resolve the transportation problems that plague the cities today. However, because of the increasing lead time necessary for development and application of technological innovation, it will be desirable to allocate appreciable funds for design, development, and demonstration of new hardware systems. Because of the current public concern with safety and pollution, coming years should see substantial efforts by both industry and government toward the development of totally new technological systems. Such systems will require major changes in urban structure, and because of their high costs, their development will be gradual and based on market demand rather than as welfare investment. Since there will be little immediate market, the government will have to underwrite the majority of the costs, as is done with space systems, for example.

Refining data needs and information retrieval systems—The availability of high-capacity electronic data processing systems has permitted the geometric increase in the availability of data for transportation planning and research purposes. The next few years will continue to see further application of computers in planning and research and in design, maintenance, construction, and operation of transportation facilities as well. Much progress needs to be made in computer application, particularly toward integrated processing of management, planning, and design information. Standardization of input, output, and software is badly needed.

If the Highway Research Board is to play an important role in the solution of problems of this magnitude, it will not have time to reflect on past glories. After a short pause to acknowledge the passing of 50 years, the Highway Research Board will move right along into the future.

The Executive Committee of the Highway Research Board is made up of prominent individuals from federal and state governments, industry, and educational institutions. Some of the members of the 1967 Executive Committee are pictured here, along with key members of the HRB staff.

Appendixes

APPENDIX A *An Act to Incorporate the National Academy of Sciences*

THIRTY-SEVENTH CONGRESS OF THE UNITED STATES OF AMERICA AT THE THIRD SESSION

An ACT To incorporate the National Academy of Sciences

Be it enacted by the Senate and House of Representatives of the United States of America in Congress assembled, That Louis Agassiz, Massachusetts; J. H. Alexander, Maryland; S. Alexander, New Jersey; A. D. Bache, at large; F. B. Barnard,[1] at large; J. G. Barnard, United States Army, Massachusetts; W. H. C. Bartlett, United States Military Academy, Missouri; U. A. Boyden,[2] Massachusetts; Alexis Caswell, Rhode Island; William Chauvenet, Missouri; J. H. C. Coffin, United States Naval Academy, Maine; J. A. Dahlgren,[2] United States Navy, Pennsylvania; J. D. Dana, Connecticut; Charles H. Davis, United States Navy, Massachusetts; George Engelmann, Saint Louis, Missouri; J. F. Frazer, Pennsylvania; Wolcott Gibbs, New York; J. M. Gilless,[3] United States Navy, District of Columbia; A. A. Gould, Massachusetts; B. A. Gould, Massachusetts; Asa Gray, Massachusetts; A. Guyot, New Jersey; James Hall, New York; Joseph Henry, at large; J. E. Hilgard, at large, Illinois; Edward Hitchcock, Massachusetts; J. S. Hubbard, United States Naval Observatory, Connecticut; A. A. Humphreys, United States Army, Pennsylvania; J. L. Le Conte, United States Army, Pennsylvania; J. Leidy, Pennsylvania; J. P. Lesley, Pennsylvania; M. F. Longstreth, Pennsylvania; D. H. Mahan, United States

Military Academy, Virginia; J. S. Newberry, Ohio; H. A. Newton, Connecticut; Benjamin Peirce, Massachusetts; John Rodgers, United States Navy, Indiana; Fairman Rogers, Pennsylvania; R. E. Rogers, Pennsylvania; W. B. Rogers, Massachusetts; L. M. Rutherfurd, New York; Joseph Saxton, at large; Benjamin Silliman, Connecticut; Benjamin Silliman, Junior, Connecticut; Theodore Strong, New Jersey; John Torrey, New York; J. G. Totten, United States Army, Connecticut; Joseph Winlock, United States Nautical Almanac, Kentucky; Jeffries Wyman, Massachusetts; J. D. Whitney, California; their associates and successors duly chosen, are hereby incorporated, constituted, and declared to be a body corporate, by the name of the National Academy of Sciences.

SEC. 2. *And be it further enacted,* That the National Academy of Sciences shall consist of not more than fifty ordinary members, and the said corporation hereby constituted shall have power to make its own organization, including its constitution, by laws, and rules and regulations; to fill all vacancies created by death, resignation, or otherwise; to provide for the election of foreign and domestic members, the division into classes, and all other matters needful or usual in such institution, and to report the same to Congress.

SEC. 3. *And be it further enacted,* That the National Academy of Sciences shall hold an annual meeting at such place in the United States as may be designated, and the Academy shall, whenever called upon by any department of the Government, investigate, examine, experiment, and report upon any subject of science or art, the actual expense of such investigations, examinations, experiments, and reports to be paid from appropriations which may be made for the purpose, but the Academy shall receive no compensation whatever for any services to the Government of the United States.

GALUSHA A. GROW,
Speaker of the House of Representatives.
SOLOMON FOOT,
President of the Senate pro tempore.

Approved, March 3, 1863.
ABRAHAM LINCOLN, *President.*

[1] The correct name of this charter member was F. A. P. Barnard.
[2] Declined.
[3] The correct name of this charter member was J. M. Gilliss.

Amendments

AN ACT To amend the act to incorporate the National Academy of Sciences

Be it enacted by the Senate and House of Representatives of the United States of America in Congress assembled, That the act to incorporate the National Academy of Sciences, approved March third, eighteen hundred and sixty-three, be, and the same is hereby, so amended as to remove the limitation of the number of ordinary members of said Academy as provided in said act.

Approved, July 14, 1870.

AN ACT To authorize the National Academy of Sciences to receive and hold trust funds for the promotion of sciences, and for other purposes

Be it enacted by the Senate and House of Representatives of the United States of America in Congress assembled, That the National Academy of Sciences, incorporated by the act of Congress approved March third, eighteen hundred and sixty-three, and its several supplements be, and the same is hereby, authorized and empowered to receive bequests and donations and hold the same in trust, to be applied by the said Academy in aid of scientific investigations and according to the will of the donors.

Approved, June 20, 1884.

AN ACT To amend the act authorizing the National Academy of Sciences to receive and hold trust funds for the promotion of science, and for other purposes

Be it enacted by the Senate and House of Representatives of the United States of America in Congress assembled, That the act to authorize the National Academy of Sciences to receive and hold trust funds for the promotion of science, and for other purposes, approved June twentieth, eighteen hundred and eighty-four, be, and the same is hereby, amended to read as follows:

"That the National Academy of Sciences, incorporated by the act of Congress approved March third, eighteen hundred and sixty-three, be, and the same is hereby, authorized and empowered to receive by devise, bequest, donation, or otherwise, either real or personal property, and to hold the same absolutely or in trust, and to invest, reinvest, and manage the same in accordance with the provisions of its constitution, and to apply said property and the income arising therefrom to the objects of its creation and according to the instructions of the donors: *Provided, however,* That the Congress may at any time limit the amount of real estate which may be acquired and the length of time the same may be held by said National Academy of Sciences.

SEC. 2. That the right to alter, amend, or repeal this act is hereby expressly reserved.

Approved, May 27, 1914.

APPENDIX B *Executive Order*

The National Research Council was organized in 1916 at the request of the President by the National Academy of Sciences, under its congressional charter, as a measure of national preparedness. The work accomplished by the Council in organizing research and in securing cooperation of military and civilian agencies in the solution of military problems demonstrates its capacity for larger service. The National Academy of Sciences is therefore requested to perpetuate the National Research Council, the duties of which shall be as follows:

1. In general, to stimulate research in the mathematical, physical and biological sciences, and in the application of these sciences to engineering, agriculture,

medicine and other useful arts, with the object of increasing knowledge, of strengthening the national defense, and of contributing in other ways to the public welfare.

2. To survey the larger possibilities of science, to formulate comprehensive projects of research, and to develop effective means of utilizing the scientific and technical resources of the country for dealing with these projects.

3. To promote cooperation in research, at home and abroad, in order to secure concentration of effort, minimize duplication, and stimulate progress; but in all cooperative undertakings to give encouragement to individual initiative, as fundamentally important to the advancement of science.

4. To serve as a means of bringing American and foreign investigators into active cooperation with the scientific and technical services of the War and Navy Departments and with those of the civil branches of the Government.

5. To direct the attention of scientific and technical investigators to the present importance of military and industrial problems in connection with the war, and to aid in the solution of these problems by organizing specific researches.

6. To gather and collate scientific and technical information, at home and abroad, in cooperation with governmental and other agencies, and to render such information available to duly accredited persons.

Effective prosecution of the Council's work requires the cordial collaboration of the scientific and technical branches of the Government, both military and civil. To this end representatives of the Government, upon the nomination of the National Academy of Sciences, will be designated by the President as members of the Council, as heretofore, and the heads of the departments immediately concerned will continue to cooperate in every way that may be required.

(Signed) WOODROW WILSON.

The White House
11 May, 1918
(No. 2859)

Executive Order
Amendment of Executive Order No. 2859
of May 11, 1918, Relating to
the National Research Council

Executive Order No. 2859 of May 11, 1918, relating to the National Research Council, is hereby amended to read as follows:

"NATIONAL RESEARCH COUNCIL OF THE NATIONAL ACADEMY OF SCIENCES

"WHEREAS the National Research Council (hereinafter referred to as the Council) was organized in 1916 at the request of the President by the National Academy of Sciences, under its congressional charter, as a measure of national preparedness; and

"WHEREAS in recognition of the work accomplished by the National Academy of Sciences through the Council in organizing research, in furthering science, and in securing cooperation of government and non-government agencies in the solution of their problems, the Council has been perpetuated by the Academy as requested by the President in Executive Order No. 2859 of May 11, 1918; and

"WHEREAS the effective prosecution of the Council's work requires the close cooperation of the scientific and technical branches of the Government, both military and civil, and makes representation of the Government on the Council desirable:

"NOW, THEREFORE, by virtue of the authority vested in me as President of the United States, it is ordered as follows:

"1. The functions of the Council shall be as follows:

"(a) In general, to stimulate research in the mathematical, physical, and biological sciences, and in the application of these sciences to engineering, agriculture, medicine, and other useful arts, with the object of increasing knowledge, of strengthening the national defense, and of contributing in other ways to the public welfare.

"(b) To survey the broad possibilities of science, to formulate comprehensive projects of research, and to develop effective means of utilizing the scientific and technical resources of the country for dealing with such projects.

"(c) To promote cooperation in research, at home and abroad, in order to secure concentration of effort, minimize duplication, and stimulate progress; but in all cooperative undertakings to give encouragement to individual initiative, as fundamentally important to the advancement of science.

"(d) To serve as a means of bringing American and foreign investigators into active cooperation with the scientific and technical services of the Department of Defense and of the civil branches of the Government.

"(e) To direct the attention of scientific and technical investigators to the importance of military and industrial problems in connection with national defense, and to aid in the solution of these problems by organizing specific researches.

"(f) To gather and collate scientific and technical information, at home and abroad, in cooperation with governmental and other agencies, and to render such information available to duly accredited persons.

"2. The Government shall be represented on the Council by members who are officers or employees of specified departments and agencies of the executive branch of the Government. The National Academy of Sciences shall specify, from time to time, the departments and agencies from which Government members shall be designated, and shall determine, from time to time, the number of Government members who shall be designated from each such department and agency. The head of each such specified department or agency shall designate the officers and employees from his department or agency, in such numbers as the National Academy of Sciences shall determine, who shall be members of the Council, but shall designate only those persons who are acceptable to the Academy."

This order shall not be construed as terminating the tenure of any person who has heretofore been designated as a member of the Council.

/signed/DWIGHT D. EISENHOWER

The White House
May 10, 1956
(No. 10668)

APPENDIX C *Report to the Chairman of the Engineering Division from the Division Representatives Attending the Conference in Chicago, October 8, 1919*

IMPORTANCE: (a) The nation and the states are starting to spend billions of dollars within the next 15 to 20 years on highway improvement. Already the Federal Congress has appropriated $275,000,000 for federal aid in this work, at the rate of $100,000,000 per year for 1919-20. (b) There is urgent need for very extensive immediate scientific highway research to establish the fundamental data needed by highway engineers in designing and constructing the highways which are to be built.

HIGHWAY RESEARCH AGENCIES IN THE UNITED STATES may be enumerated as follows:

(a) The United States Bureau of Public Roads is spending a sum estimated by Mr. Marston as probably $150,000 per year in scientific research as distinguished from actual road construction, and Mr. Marston believes that the activity of the Bureau of Public Roads in this direction should be increased rather than diminished. Extensive scientific work in measuring stresses in pavement slabs from motor truck loads, the effects of impact and many other scientific tests are already under way.

(b) Many of the strong engineering colleges have excellent testing laboratories, and there are a few engineering experiment stations organized and engaged constantly in scientific engineering research. Many of these laboratories might engage in highway research if work were assigned and funds provided.

As illustrating such work, it may be said that the Iowa Engineering Experiment Station is required by state law to spend at least $10,000 annually of its funds for road experimentation, and that it is engaged in research on the economic theory of highway improvement involving the determination of all elements of cost of highway transportation and the effect upon each element of cost of transportation of each element of road improvement, upon which investigation it proposes to spend at least $7,500 per year for a period of several years.

(c) Practically every state now has a state highway department. The more completely developed of those departments are already undertaking highway research and should be effective agencies in a national program. The state highway departments have a national association [American Association of State Highway Officials] so that they can act unitedly, and they have a Committee on Investigations and Tests, of which Professor T. R. Agg, Ames, Iowa, is chairman.

(d) Municipal testing laboratories are available in a number of large cities which could do valuable work.

(e) Manufacturers' research departments and associations should be able to take a very valuable part in a national research program.

(f) Commerical laboratories.

(g) Technical societies. The great national engineering societies can render effective assistance in organizing a national program and in determining the use to be made of the scientific data to be secured.

(h) Consulting highway engineers.

NEED FOR COORDINATION OF HIGHWAY RESEARCH AGENCIES: The number and diversity of the highway research agencies available in the United States make it apparent that their work will be more or less desultory in character and involve much unnecessary duplication, and will probably omit important researches unless they can be coordinated in a national program of highway research. This coordination should be undertaken at once. The National Research Council seems to have been established for doing just this sort of service.

PLANS FOR A NATIONAL PROGRAM OF HIGHWAY RESEARCH: The undersigned recommended to the Engineering Division of the National Research Council that it undertake at once the inauguration of a national program of highway research.

Since a number of research committees will be required working on this subject, we recommend the appointment of a subcommittee of the Engineering Division to cooperate with the chairman in coordinating the various highway research committees.

We recommend that the Engineering Division of the National Research Council proceed at once to establish six highway research committees as follows:

(1) *Committee on the Economic Theory of Highway Improvement*:

Purpose: Determination of all elements of cost of highway transportation (both motor vehicle and animal drawn), and the effect upon each cost element of each feature of highway improvement (improved surface, grade reduction, elimination of rise and fall, etc.) to enable reliable and scientific determination of the sums which could be economically expended for specific highway improvements.

Agencies: (a) The Iowa Engineering Experiment Station (continuing work already in progress).

(b) The U. S. Bureau of Public Roads (continuing work already in progress).

(c) State Highway Departments.

(d) Motor Vehicle Manufacturers.

Resources: (a) Seven thousand five hundred dollars annually from the Iowa Engineering Experiment Station (promised by the Director).

(b) Probably a material sum annually from the U.S. Bureau of Public Roads.

(c) Services of employees and apparatus of State Highway Departments and Municipal Departments.

(d) Contributions of materials, apparatus and funds by Motor Vehicle Manufacturers.

(2) *Committee on Structural Design of Roads*:

Purpose: Experimental determination of stresses, strains and behavior, of impact and abrasion in actual roadways under actual traffic; including also the effect of temperature variations, freezing and thawing, non-uniform sub-grades and other factors; in order to establish reliable scientific engineering theory for the structural design of roads.

Agencies: (a) U. S. Bureau of Public Roads (continuing extensive experiments already in progress).

(b) Such other technical laboratories as can be enlisted.

(c) The Society of Automotive Engineers.

(d) The National Automobile Chamber of Commerce.

(e) The United States War Department.

(f) Manufacturers of motor vehicles and accessories.

Resources: (a) The U. S. Bureau of Public Roads is already expending a large amount annually for this work and it is expected that it will continue adequate support for the work in the future.

(b) The State Highway Departments will undoubtedly contribute services of employees and the use of laboratories to an extent representing an annual expenditure of a very large sum of money.

(c) It is hoped that manufacturers will continue to donate materials.

(d) Any actual road construction needed in various sections of the country can be provided through Federal and State cooperation from road funds already established.

(3) *Committee on Properties of Road Materials*:

Purpose: To investigate the properties of road materials and their combinations, including devising and standardizing tests.

Agencies: Government, College, Highway Department, Municipal, Commercial and Manufacturers' laboratories which might be enlisted.

Resources: Special assignments of men, laboratories and funds by government departments, State Highway Departments and the institutions and organizations supporting the various laboratories enlisted.

(4) *Committee on Methods of Road Construction*: The highway engineers present at the conference urgently recommended the establishment of this committee to meet their needs.

Purpose: To collect and disseminate information on the various methods of securing efficient results in road construction, under various conditions and in the different parts of the country. [From paper *A National Program for Highway Research*, presented at Annual Meeting of AASHO, December 8-11, 1919, by Dean Anson Marston, Iowa State Highway Commission, and published in *Public Roads*, Vol. 2, Nos. 21-22, January-February 1920, pp. 34-37.]

(5) *Committee on Maintenance Methods and Design:* The highway engineers present at the conference urgently requested this committee also.

Purpose: To collect and disseminate information on the various maintenance methods and principles of maintenance design which are found to give best results under different conditions in the various sections of the country. [See reference above.]

(6) *Committee on Highway Bridge and Culvert Research*: Bridge Departments of the State Highway Departments urge the need for scientific researches to determine many factors of highway bridge and culvert design.

Purpose: Investigation of the waterways required for different classes of culvert and drainage areas. Investigation of the distribution of wheel loads over bridge floors. Other unsolved problems of highway culverts and bridges. [See reference under (4) above.]

In conclusion, the undersigned wish to say that they feel that it is quite important that National Research Council should secure at once the active cooperation of the U. S. Bureau of Public Roads and of the various State Highway Departments in inaugurating a national program of highway research. To this end, we recommend that the executive committee of the division take steps to secure if possible, by appointment of the President of the United States, the Chief of the Bureau of Public Roads as a member at large of the Engineering Division in accordance with Article 5, Section 3, of the Organization Pamphlet (or constitution) of the National Research Council.

Respectfully submitted,
A. N. Talbot
H. H. Porter
G. S. Webster
Anson Marston

October 27, 1919

APPENDIX D

A More Definite Statement of the Highway Research Work

It is an accepted theory that highways and their use are one problem and that this form of transportation is vital to the welfare and prosperity of the nation. Today it is an undeveloped subject. This committee is seeking fundamental information.

AT the recent conference of educators and industrial leaders called by the Highway and Highway Transport Education Committee, Prof. W. K. Hatt, director of the Highway Research Committee of the Division of Engineering, National Research Council, read a more definite statement of the work planned by his committee than has heretofore been made public. This report was referred to in the recently printed account of that meeting. The research work undertaken by this committee dovetails completely with the forward plans of the Education Committee, as the research must develop the information necessary for the proper education of the highway and highway transport engineer of the future.

The outline of the work undertaken by Prof. Hatt and his associates is still subject to change. The tentative outline presented at the meeting referred to follows. The committee invites suggestions.

By W. K. Hatt

INDIVIDUALS in industry who have endeavored to state the problem in approximate statistics tell us that there are 10 billions of dollars invested in self propelled vehicles, and that the turnover is three billions annually. That there are ten passenger vehicles to one freight vehicle, and that the problems of speed and safety are most important. That the annual expenditure for operation of vehicles is twelve times the annual expenditure on the roads; therefore the field of research on cost of vehicle operation arising from the road surface and from the vehicle itself, must be kept in the foreground. That transportation by self propelled vehicles is the most expensive of all commercial forms; it will increase. Everyone pays for efficiency.

For the purpose of a coordinated and comprehensive program of Highway Research, the writer has been endeavoring to bring into the picture all the elements of the situation in Highway Transport: Engineering, including vehicle and road; Economics of Transportation; Administration; Finance. Some comprehensive and logical assemblage will be helpful.

It appears that there are many dimly seen figures which should be advanced from the background; there is much that is unknown. We may ask a few questions which cannot be completely answered.

Fundamental Questions in Highway Transport

The Transport Unit

(1) What is the economical highway track unit for each of the several situations, e.g., intercity, farm to market?
What is the cost of transport arising from vehicle and from road?

(2) What is the relation of this economical unit to other systems of transport, e.g., electric and steam, in a unified system?

(3) To what extent, as a matter of public policy, should any transport unit be indirectly subsidized?

(4) What traffic regulations should be imposed on such economical unit over other types of road?
What fees should be charged for service rendered to vehicle by the road?

(5) What should be the proportion of the total traffic supplied by such economical unit to justify a special design of road for such unit?

(6) What prediction can be made of future changes in general traffic and what is the influence of these on the economics of the present situation?

(7) How should passenger traffic over the highway be evaluated?

The Road

(1) What type of road paving should be selected for a specified transport unit?

(2) If the road cannot be economically fitted to the truck transport unit, can the latter be modified in design to fit the road?

(3) How should the design of the road and paving be modified to meet changing conditions of subgrade, climate, etc.? How shall sub-soils be improved?

(4) What sum of money is the locating engineer justified in spending to avoid increase in distance, curvature, rise and fall, maximum grade, maximum curve?

(5) What system of maintenance and organization is best fitted for types of roads, differing in traffic, in materials, and in climate?

(6) What is capacity of a road of given width for type of vehicle as expressed in vehicles per hour, ton-miles per year, etc.? What is the appropriate unit for expressing traffic for various purposes?

(7) (In construction many questions arise in selection, production and economical use of materials, standardization and regulation.)

(8) How can the volumetric changes in roads be overcome?

(9) What is the economical life of various types of roads—that is, when maintenance charges exceed earning value?

Administration

(1) What should be the policy in control of truck and bus transportation systems, terminals, routing, etc.?

(2) What police regulations should control use of roads?

(3) What is the best administrative and executive organization for administration and operation of roads?

(4) What principles should govern the selection of a system of roads in its various parts, as influenced by inter-state, intra-state, county, local traffic, etc.?

Financing

(1) What should be method of financing construction and maintenance of roads? What portions of cost from long term bonds, and what from current funds? What form of bonds should be issued and how create a market for them?

(2) What should be the relation between life of bonds and economical life of road?

(3) To what extent do social betterment, military use, i.e., social value, and other imponderables enter into highway policy?

(4) What should be the distribution of costs as between Federal, State, county, township, property benefited, the user and other units?

(5) How shall the future maintenance charges on completed road systems be met? Shall the user pay all of these?

(6) How shall safety be ensured on the roads?

Answers to these questions cannot be made without data that are at present unavailable.

Research is necessary, and a mobilization of the efforts, of research agencies in a comprehensive program. The Highway Research Committee of the Division of Engineering of the National Research Council has undertaken the coordination of such research. The National Research Council will not engage in research directly.

The chart reproduced herewith is devised to indicate the field of research as divided into subfields in which research should be developed, and from which data should come to enable answers to be made to these questions among others.

Some of the studies that should be made are as follows:

(1) To develop a traffic census blank. Here a traffic classification must be made, the purpose of the census determined, and the various forms and instructions standardized.
(2) *In order to determine the cost of transport* a statistical table must be made that notes all of the elements of cost; sometimes only a few of these are reported.
(3) To study the operating costs of elements entering into location of highways, such as distance, grade, curvature.
(4) To study loads on roads as produced by the vehicle.
(5) To study design of vehicles with a view to lessening their effects on the road.
(6) To study supporting power and improvement of subgrades and the relations to design of paving.
(7) To study resistance of concrete slabs to alternate stresses and to surface loads.
(8) To study proportioning and use of bituminous materials.
(9) To study bonding of brick surfaces.
(10) To study volume changes and the means of meeting them.
(11) To study operations of concrete mixers.
(12) To study the organization and economics of construction plants.
(13) To study sand-clay, top-soil and gravel roads.
(14) To study cellular and other new types of paving.

There is apparently a widespread activity in highway research throughout the United States on the part of the Bureau of Roads, the U. S. Army, the State Highway Commissions, the universities and of industrial organizations and an earnest desire to put highway construction on a scientific basis.

The economical features are under critical examination by such organizations as the National Chamber of Commerce.

We should be able to express quantitatively the results of a standardized economic survey of a road project, just as in the case of a water-power project for instance, except for those imponderables, which, like social betterment and public policy, influence the conclusions so profoundly.

It is not too much to say that the situation is critical and that the sooner those interested come to a basis of fact the more assurance we will have that the public will not interrupt progress in providing for highway transport because of a general feeling of insecurity.

TENTATIVE CHART
TO SHOW ORIGIN OF FLOW OF DATA FROM

THE FIELD OF HIGHWAY RESEARCH

IMPORTANT ELEMENTS DEMANDING STUDY ARISING IN THE DIVISIONS OF I TO V
(STRUCTURES OMITTED) (ELEMENTS ARE OFTEN INTER-RELATED)
STATIC CHART. COMMITTEES ORGANIZED TO OCCUPY FIELD

I ECONOMICS (DATA FOR ECONOMIC SURVEY AND STUDY OF PROJECT)	II OPERATION	III DESIGN (ROAD)	IV DESIGN (VEHICLES) AS RELATED TO ROAD	V CONSTRUCTION
1. TRAFFIC STUDIES (REGIONAL) A. DISTRIBUTION IN REGION B. CHARACTER (TRAFFIC BLANK) VEHICLE WEIGHT AND DISTRIBUTION SPEED TIRE (CONDITION) COMMODITY LENGTH HAUL C. METHOD OF EXPRESSING UNIT OF TRAFFIC D. PREDICTED CHANGES E. OTHER TRAFFIC ON STEAM AND ELECTRIC ROADS F. CENTRAL SOURCES OF TRAFFIC 2. COMMUNITY NEEDS SYSTEMS OF ROADS IN CLASSES FOR INDUSTRIES, ETC. INTANGIBLES 3. COST OF TRANSPORT CAPITAL COST ROAD VEHICLE FIXED CHARGES OVERHEAD OPERATION MAINTENANCE ROUTINE REPLACEMENT ON ROAD ON VEHICLE EQUIVALENT UNITS X-AUTO = Y-TON TRUCK, ETC. ECONOMIC LIFE 4. ECONOMICS OF LOCATION COST OF DISTANCE RISE AND FALL CURVATURE RULING GRADE RULING CURVE 5. FINANCING BONDS TAXES FEES ETC. 6. HIGHWAY VALUATION INCREMENT OF LAND VALUES	1. CONTROL OF TRAFFIC ROUTING TERMINALS FRANCHISES POLICE REGULATIONS 2. ACCIDENT INSURANCE 3. PLANNING SYSTEMS OF TRANSPORT FINANCING ENVIRONMENT RELATION TO OTHER TRANSPORT ORGANIZATION 4. FINANCING (SEE I) BONDS, TAXES, FEES 5. DISTRIBUTION OF COSTS TRAFFIC PROPERTY POLITICAL UNITS 6. MAINTENANCE SYSTEMS 7. MAINTENANCE MACHINERY 8. MAINTENANCE METHODS ROUTINE REPLACEMENT SNOW REMOVAL, ETC. 9. TRAIL MARKING 10. COST ACCOUNTING 11. SAFETY	1. SUBSOIL STUDIES PROPERTIES PHYSICAL MECHANICAL CHEMICAL DRAINAGE SUPPORTING POWER IMPROVEMENT BY TREATMENT EFFECT OF ROAD DEFORMATIONS EFFECT OF CLIMATE DISTRIBUTION OF PRESSURE 2. BASE COURSE CHARACTER, TYPE THICKNESS MATERIALS CROSS SECTION 3. SURFACE CHARACTER THICKNESS MATERIALS CROSS SECTION WEAR BY TRAFFIC WEAR BY ELEMENTS IMPACT OF TRAFFIC TRACTIVE RESISTANCE WEAR OF TIRE WEAR OF VEHICLE DUSTING INFLUENCE OF LOCALITY 4. CROSS SECTION WIDTH CROWN SHOULDERS DITCHES 5. LOADS STATIC IMPACT SURFACE INTEGRAL AS EFFECTED BY DESIGN OF VEHICLE 6. DESIGN OF INTEGRAL SLAB STRENGTH AND STIFFNESS OF SOLID PRECAST CELLULAR 7. VOLUME CHANGES JOINTS SHOVING 8. REINFORCING THEORY OF AMOUNT KIND DISTRIBUTION DIRECTION	1. DESIGN OF VEHICLE POWER GEAR RATIO BRAKING ETC. ETC. EFFECT ON LOADS SPRUNG UNSPRUNG DISTRIBUTION ECONOMY OF OPERATION AND MAINTENANCE (SEE AUTOMOTIVE INDUSTRY) ECONOMIC LIMIT OF SIZE OF TRUCK IN VARIOUS SITUATIONS COST OF TRUCK TRANSPORT (SCHEDULE OF ELEMENTS) ECONOMIC LIMIT OF HAUL COST ACCOUNTING 3. SURFACE TRACTIVE EFFORT WEAR ON TIRES LOADS MAINTENANCE OF VEHICLE 4. ALIGNMENT CURVES (SPEED) GRADES 5. CROSS SECTION WIDTH CROWN 6. SAFETY	1. MATERIALS BITUMINOUS NON BITUMINOUS FUNDAMENTAL MECHANICAL PROPERTIES METHODS OF TEST STANDARD TESTS SPECIFICATION PREPARATION AND TREATMENT (SEE SPECIAL LIST) PROPORTIONING 2. MIXING EFFICIENCY OF MIXER CENTRAL MIXING PLANTS 3. PLACING 4. METHODS OF TESTING ROADS INSTRUMENT CORES 5. DESIGN OF EXPERIMENTAL ROADS 6. DRAINAGE (AND DRAINAGE STRUCTURES) 7. IMPACT ON BRIDGES (SEE DESIGN) 8. REINFORCING HANDLING AND PLACING 9. INSPECTION 10. PLANT-DESIGN AND CONTROL 11. COST ACCOUNTING 12. CONSTRUCTION CONTRACTS

NOTE.—This list of research studies applies to all classes of roads—brick, concrete, macadam, sand-clay, gravel and bituminous. Special problems arising in special types of roads will appear in a supplementary detailed chart.

APPENDIX E *Topical Chronology of Principal Events Relating to the Highway Research Board*

DATE	EVENT
1861-65	Civil War.
1863	National Academy of Sciences established.
1869	Steam roller developed.
1871	Portland cement concrete construction.
1872	First brick pavement laid, Charleston, West Virginia.
1873	Asphalt concrete construction.
1877	First bicycle built.
1878	District of Columbia Highway Department established.
1885-89	Development of pneumatic tire.
1888-89	First auto truck developed by Semple Clark, Pittsburgh.
1889	Sand-clay road development.
1889	European automobile exhibited by Benz at World's Fair, Paris.
1890	Census counts 62,947,714 people in United States, 20 million horses and mules.
1890	Inauguration of "Good Roads Movement" by League of American Wheelmen.
1891	First state-aid law, New Jersey.
1892	First American automobile (Duryea).
1893	First brick (rural) road.
1893	Massachusetts Highway Department established.
1893	First American portland cement concrete street, Bellefontaine, Ohio.
1893	Office of Road Inquiry established in U. S. Department of Agriculture.
1895	Four automobiles registered in United States.
1896	Rural Free Delivery established.
1897	First "object lesson" road by Office of Road Inquiry.
1898	Dust palliatives used.
1899	First auto fatality, September 13, New York City.
1900	Population reached 76 million people, 26 million horses and mules.
1900	Bureau of Standards established.
1900	About 200 miles of rural pavements in existence.
1900	12 to 15 billion horse-drawn vehicle-miles logged.
1900	8,000 autos registered.
1900-19	Experimental road tests in Tennessee, Texas, Illinois, Rhode Island, and other states.
1902	American Road Builders' Association established.
1902	American Automobile Association established.
1903	First automobile trip across United States completed.
1903	Ford Motor Company organized.
1904	New York City subway opened.

DATE	EVENT
1905	Office of Road Inquiry name changed to Bureau of Public Roads.
1905	Division of Tests organized in Bureau of Public Roads.
1906	Bituminous macadam surface tried.
1908	Private portland cement concrete road built on Long Island.
1909	Less than 50 percent of engineering colleges offer courses in highway engineering.
1909	First rural portland cement concrete road.
1909	"Peak year for travel by horse power."
1910-19	Highway research increased in land grant colleges and highway departments.
1910 Circa	Semaphore traffic signals in use.
1911-13	Bituminous macadam experimental road in Chevy Chase, Maryland.
1913	Golden Anniversary of National Academy of Sciences.
1913	First size-weight control for motor vehicles.
1914	American Association of State Highway Officials organized.
1914-18	World War I (United States entered April 6, 1917).
1915	Arlington Experimental Farm test program by Bureau of Public Roads instituted to test soils, impact tests, etc.
1915	About 500 miles of rural pavements in existence.
1915	Divided highway built in Delaware.
1916	Federal-Aid Road Act for construction of rural roads passed.
1916	National Research Council established with seven divisions of science and technology, including Division of Engineering.
1916-17	Majority of state highway departments established.
1918	World War I ends November 11.
1918	First consensus on size-weight limits; 28,000-lb gross load, 800-lb per inch of tire width arbitrary recommendation.
1918	First federal road built in California (2.4 miles).
1918	Three-color traffic signal installed in New York City.
1919	Experimental test roads begun by Bureau of Public Roads.
1919	Gasoline tax inaugurated in Oregon.
1919	Conference under aegis of Division of Engineering of National Research Council takes steps to establish National Highway Research Program, establishing the Committee on Highway Research.
1919	A national program for highway research advocated by Anson Marston at AASHO Annual Meeting.
1920	United States population climbs to 106 million.
1920	Advisory Board on Highway Research organized November 11.
1920	AASHO members number 31 state highway departments and BPR.
1920	About 20 billion vehicle-miles of travel.
1920	Fatality rate per 100 million vehicle-miles: approximately 30 for horse-drawn vehicles, approximately 20 for motor vehicles.
1920	8,132,000 autos and 1,108,000 trucks registered, and about 9 million horse-drawn vehicles being used.

DATE	EVENT
1920	3,191 miles of federal-aid roads completed.
1920	Arlington Road Test (elliptical).
1920-21	Alfred D. Flinn, Vice Chairman, Division of Engineering, Interim Director (November 11, 1920-June 30, 1921).
1920-23	Bates Road Test, Illinois (the beginning of controlled-traffic road tests).
1921	Suggested "outline of activities for highway research" (24 problems) by W. K. Hatt.
1921	Federal-Aid Highway Act strengthened and Federal-Aid System selected.
1921	AASHO welcomes HRB establishment.
1921	Traffic signal lights installed in Detroit.
1921-22	Pittsburg (California) Road Test.
1921-24	W. K. Hatt, Director of HRB (July 1, 1921-March 31, 1924).
1922	First Annual Meeting of HRB, January 16.
1922	First HRB publication (Proceedings).
1922	Second Annual Meeting of HRB in November.
1922	First highway research census in October (479 projects)
1922	Service test roads included those at Bayberry and Lancaster, Pennsylvania; Casper, Wyoming; Milwaukee, Wisconsin; Delaware; and Washington.
1922	Interlocking traffic signals developed, Houston.
1922	Ethyl gasoline marketed.
1922	Balloon tires developed for autos.
1923	Need for HRIS noted.
1923	Formula for pavement depth devised.
1924	HRB offices moved to new Academy Building, 2101 Constitution Avenue.
1924	"Contact men" appointed in 45 highway departments.
1924	Organization name changed to Highway Research Board (December 1924, effective January 1, 1925).
1924	"News Items" published during year.
1924	More than 4 million motor vehicles sold during year.
1924	"Use of horse and carriage for transport of passengers almost obsolete."
1924	High-early-strength cement developed.
1924	One auto registered for every seven persons in United States.
1924	First National Conference on Street and Highway Safety called by Herbert Hoover, Secretary of Commerce.
1924-25	First research project conducted by HRB.
1924-28	Charles M. Upham, Director from April 1, 1924, to January 20, 1928.
1925	Adoption of uniform signs and markings for U. S. Highways.
1925	Bronx River Parkway built with full access control.
1925	"Contact men" appointed at more than 100 universities.
1925	First foreign "contact man" appointed at University of Mexico.

DATE	EVENT
1925	Five sponsored investigations under way.
1925	Highway Research News inaugurated (October 1925 to March 1927).
1925	293 attend Fifth Annual Meeting.
1925-26	Summary Bulletins of Third, Fourth, Fifth, and Sixth Annual Meetings published.
1926	Board joins Permanent International Association of Road Congresses (affiliated 1926-31, resumed in 1967).
1926	Water-cement ratio concept developed by D. A. Abrams.
1926	Belt highways built around congested areas.
1927	Research Activities Committee created by AASHO.
1927	25,000 traffic fatalities; rate increasing 5 percent annually.
1928	Roy W. Crum, Director from March 1, 1928, to May 31, 1951.
1928	News Letter initiated in May (continued through March 1933).
1928	Second research census conducted.
1928	Low Cost Road Report published.
1929	Car cemeteries appeared along highways.
1929	Stock market crash on October 29; depression begins.
1929	$500,000 spent on research in 1929—half by BPR, remainder by 24 states.
1929	"A Program of Highway Research" published (separately and in Proceedings) as proposed by MacDonald and Crum (130 problems).
1929	Fred Burggraf employed as special investigator on February 15.
1929	Plan to restructure existing 11 committees into 6 broad committees ("sections") and subcommittees presented in December (effected January 1930).
1930	206 billion miles of motor vehicle travel recorded.
1930	26,750,000 vehicles registered.
1930	$2½ billion expended on highways.
1930	United States population reaches about 123 million.
1930	Institute of Traffic Engineers established.
1930	HRB reorganized into sections and committees.
1930	293 attend Annual Meeting.
1930	"HRIS" begun.
1930-36	Arlington Road Test in progress by BPR tests static loads on rigid pavements.
1931	$1 million spent on highway research, of which BPR spent $600,000.
1931	Steps taken for liaison with AASHO.
1931	Joint sponsorship (with AASHO and ARBA) of Bartlett Award initiated.
1931	54 municipal "contact men" appointed.
1932	AASHO approved 18,000-lb axle load for pneumatic tires.
1932	First joint project: HRB-AASHO on Roadside Development.
1933	National Recovery Act required landscaping of main roadsides.

DATE	EVENT
1934	G. A. Rahn suggests RCS.
1935	HRB undertakes $75,000 highway safety project for BPR.
1935	Department and committee structure set up.
1935	Department of Soils created.
1935	Reflective paints introduced.
1935	Nationwide highway planning surveys begun.
1935	First "controlled" soil-cement road built.
1936	Director made Secretary of AASHO Research Activities Committee.
1936	HRB-AASHO Highway Research Census counts 1,500 projects.
1939	Fred V. Reagel describes RCS concept.
1939-45	World War II: United States involved 1941-1945.
1940	"Public Aids to Transportation" by government.
1940	32,453,000 motor vehicles registered.
1940	580 people attend Annual Meeting.
1940	Sealed-beam headlights introduced.
1940	Pennsylvania Turnpike opened.
1940	Highway work virtually suspended because of war effort; highways declared expendable.
1940	Periodic Research Census begun (1920-40 covered).
1940	HRB Award instituted for outstanding paper.
1940	Provisions made for Associate affiliation with Board.
1940-41	HRB lecture courses in highway economics held.
1941-45	Annual Meetings held in Baltimore, St. Louis, Chicago, and Oklahoma City (1944 unassembled).
1942	Nationwide speed limit set at 35 mph to conserve fuel (gasoline rationed).
1942	Special Committee on Stimulation of Research created.
1942	Highway Research Census supplement published.
1942	Wartime Bulletins (on highway practice) inaugurated (later called Current Road Problems).
1943-54	Flexible Pavement Study conducted by HRB-BPR (Hybla Valley Project).
1944	Federal-Aid Act passed.
1945	VE Day, Japan surrendered August 14.
1945	RCS inaugurated.
1945	RCS Circulars started.
1945	Economics, Finance and Administration Department established by merging two other departments.
1945	Executive Secretary of AASHO made ex officio member of HRB Executive Committee.
1946	Air-entrained concrete appears.
1946	War restrictions removed.
1946	Digital electronic computer (ENIAC) built for military at Aberdeen, Maryland.

DATE	EVENT
1946	First President's Highway Traffic Safety Conference held.
1946	National Transportation Inquiry conducted (Board of Investigation and Research).
1946	HRB Annual Reports inaugurated.
1946	Problem census and priorities (810 problems) enumerated.
1947	Driver education offered.
1948	Tubeless tires introduced.
1948	HRB Distinguished Service Award (now Roy W. Crum Award) instituted.
1948	Highway Research Review series begun.
1948	RCS Advisory Committee instituted.
1948	Procedures established for HRB administration of AASHO or states' research.
1949	Prestressed concrete brought to United States.
1949	Colorado Highway Department organization study made.
1949-52	Highway Research Review published (6 issues).
1949-52	Road Test One–MD conducted.
1950	United States population reaches 151 million.
1950	49 million vehicles registered; 458 billion vehicle-miles traveled.
1950	$4 billion spent on highways.
1950	860 attend Annual Meeting.
1950	Plans made for cooperation among HRB and AASHO committees.
1950	Loadometer-cost-condition studies fostered by BPR.
1950-53	Korean War erupts.
1951	100 millionth car assembled.
1951	R. W. Crum succeeded by Fred Burggraf.
1951-54	WASHO Road Test.
1952	Skid-test road built in California.
1952	Detroit Area Traffic Study begun.
1953	General Motors essay contest on highway finance conducted.
1953	85 percent of passengers carried by auto.
1953	Committtee on Highway Laws established in EF&A Department.
1954	New York Thruway opened (first section).
1954	Clay Committee Report presented.
1954	Chicago Area Transportation Study begun.
1954	President's Committee for Highway Safety established.
1954	Ad Hoc Committee on Urbanization established in NAS-NRC on March 5 (became Department of Urban Transportation Planning in HRB).
1955-62	AASHO Road Test carried out.
1956	Federal-Aid Highway Act passed establishing Interstate System.
1956	Executive Order 10668 by President Eisenhower outlined functions of NRC and authorized cooperation by federal agencies.
1956	Highway fatalities number 1,168,075 since 1899.
1956	1,524 people attend Annual Meeting; first time at Sheraton-Park Hotel.

DATE	EVENT
1958	Sagamore Conference held.
1958	Piggyback rail transportation inaugurated.
1959	Compact car marketed.
1960	United States population numbers 180 million.
1960	74 million motor vehicles registered.
1960	720 billion vehicle-miles traveled, nearly 50 percent in urban areas.
1960	Thin bituminous plant-mixed, machine-placed overlays instituted.
1960	9,100 miles of Interstate System completed.
1960	2,403 people attend Annual Meeting.
1960	Wood's Hole Conference on National Transportation Problems held in August.
1960	Highway Laws Committee given Special Committee status.
1961	Fatal highway traffic accident rate 5.2 in United States.
1961	Industry Dinner held to enlist industry support.
1962	National Cooperative Highway Research Program established.
1962	HRB Director made Secretary of AASHO Evaluations Committee.
1962	AASHO Road Test Conference held in St. Louis.
1962-63	"Research in Progress" published in Abstracts.
1962-64	"Urban Transportation Research Digest" published.
1963	New motor-vehicle size and weight policy based on AASHO Road Test promulgated.
1963	Department of Urban Transportation Planning organized from Special Committee on Urban Research.
1963	NAS Centennial celebrated.
1964	D. Grant Mickle succeeded Fred Burggraf as Director in January.
1964	Highway Research Information Service began 3-year developmental period.
1964	Department of Legal Studies organized from Special Committee on Highway Laws.
1964	National Academy of Engineering established.
1965	U. S. Department of Housing and Urban Development created.
1965	Highway Beautification Act passed.
1965	Special Committee on International Cooperative Activities established.
1965	Special Committee on Long-Range Planning appointed.
1966	U. S. Department of Transportation authorized.
1966	NCHRP project "Research Needs in Highway Transportation" contracted.
1966	NRC Division of Engineering and Industrial Research resumed name of Division of Engineering.
1966	William N. Carey, Jr., succeeded D. Grant Mickle as Executive Director, October 1.
1966	HRIS reported 3,000 research projects, average annual cost of $20,000 each.
1967	HRIS became operational (July 1).
1967	Long-Range Planning Committee proposed broadened HRB scope.

DATE	EVENT
1967	HRB moved to Joseph Henry Building, 2100 Pennsylvania Avenue, N.W.
1968	Conference on Reorganization of HRB held at Airlie House in May.
1970	Reorganization of Board became effective February 1.
1970	Golden Anniversary Year.

APPENDIX F *Chairmen, Directors, and Assistant Directors*

YEAR	CHAIRMAN	DIRECTOR	ASSISTANT DIRECTOR
1920[1]	Anson Marston[2]	Alfred D. Flinn[3]	
1921	Anson Marston	Alfred D. Flinn[3]	
1922	Anson Marston	W. K. Hatt	
1923	A. N. Johnson	W. K. Hatt	E. R. Olbrich
1924	A. N. Johnson	Charles M. Upham	H. F. Janda
1925	A. N. Johnson	Charles M. Upham	H. F. Janda
1926	A. N. Johnson	Charles M. Upham	S. S. Steinberg
1927	Thomas R. Agg	Charles M. Upham	S. S. Steinberg
1928	Frank H. Eno	Roy W. Crum	
1929	Frank H. Eno	Roy W. Crum	
1930	H. S. Mattimore	Roy W. Crum	
1931	H. S. Mattimore	Roy W. Crum	
1932	George E. Hamlin	Roy W. Crum	
1933	George E. Hamlin	Roy W. Crum	
1934	Albert T. Goldbeck	Roy W. Crum	
1935	Albert T. Goldbeck	Roy W. Crum	
1936	H. C. Dickinson	Roy W. Crum	
1937	H. C. Dickinson	Roy W. Crum	
1938	Burton W. Marsh	Roy W. Crum	
1939	Burton W. Marsh	Roy W. Crum	
1940	W. W. Mack	Roy W. Crum	
1941	W. W. Mack	Roy W. Crum	Fred Burggraf
1942	Fred C. Lang	Roy W. Crum	Fred Burggraf
1943	Fred C. Lang	Roy W. Crum	Fred Burggraf
1944	Stanton Walker	Roy W. Crum	Fred Burggraf
1945	Stanton Walker	Roy W. Crum	Fred Burggraf[4]
1946	Roger L. Morrison	Roy W. Crum	Fred Burggraf[4]
1947	Roger L. Morrison	Roy W. Crum	Fred Burggraf[4]
1948	F. V. Reagel	Roy W. Crum	Fred Burggraf[4]
1949	F. V. Reagel	Roy W. Crum	Fred Burggraf[4]
1950	Ralph A. Moyer	Roy W. Crum	Fred Burggraf[4]
1951	Ralph A. Moyer	Fred Burggraf	
1952	R. H. Baldock	Fred Burggraf	
1953	R. H. Baldock	Fred Burggraf	

YEAR	CHAIRMAN	DIRECTOR	ASSISTANT DIRECTOR
1954	Walter H. Root	Fred Burggraf	Elmer M. Ward
	G. Donald Kennedy	Fred Burggraf	Elmer M. Ward
1955	G. Donald Kennedy	Fred Burggraf	Elmer M. Ward
1956	Kenneth B. Woods	Fred Burggraf	Elmer M. Ward
1957	Rex M. Whitton	Fred Burggraf	Elmer M. Ward
1958	Charles H. Scholer	Fred Burggraf	Elmer M. Ward
1959	Harmer E. Davis	Fred Burggraf	Elmer M. Ward
1960	Pyke Johnson	Fred Burggraf	Elmer M. Ward
1961	W. A. Bugge	Fred Burggraf	Elmer M. Ward
1962	R. R. Bartelsmeyer	Fred Burggraf	W. N. Carey, Jr.
1963	C. D. Curtiss	Fred Burggraf	W. N. Carey, Jr.
1964	Wilbur S. Smith	D. Grant Mickle	W. N. Carey, Jr.[5]
1965	Donald S. Berry	D. Grant Mickle	W. N. Carey, Jr.[5]
1966	J. Burch McMorran	D. Grant Mickle	W. N. Carey, Jr.[5]
1967	Edward G. Wetzel	W. N. Carey, Jr.	—[6]
1968	David H. Stevens	W. N. Carey, Jr.	—[6]
1969	Oscar T. Marzke	W. N. Carey, Jr.	—[6]
1970	D. Grant Mickle	W. N. Carey, Jr.	—[6]

[1] From November 11, 1920.
[2] Temporary Chairman.
[3] Interim Director.
[4] Associate Director.
[5] Deputy Executive Director.
[6] Effective November 1, 1966, four positions with the rank of Assistant Director were created. Appointed to the positions were James C. Allen as Assistant Director for Administration, Roy C. Edgerton as Assistant Director for Technical Activities, Paul E. Irick as Assistant Director for Special Projects, and William A. Goodwin as Program Director for NCHRP. On September 1, 1967, Krieger W. Henderson, Jr., succeeded Mr. Goodwin. The other Assistant Directors have served through 1970.

APPENDIX G *Highway Research Board Bylaws*

APPROVED APRIL 27, 1969

1. The Highway Research Board

The Highway Research Board is an organization having the purpose and scope defined below. It is a unit of the Division of Engineering of the National Research Council. It comprises an Executive Committee, a staff and an affiliated membership associated either by supporting memberships defined below or by membership on committees.

2. Authority

The Highway Research Board operates under the corporate authority of the National Academy of Sciences as a part of the Division of Engineering of the National Research Council, serving both the National Academy of Sciences and the National Academy of Engineering.

3. *Purpose*

The purpose of the Board is to advance knowledge concerning the nature and performance of transportation systems, through the stimulation of research and dissemination of information derived therefrom.

4. *Scope*

The Board will give attention to all factors pertinent to the understanding, devising and functioning of highways and urban transportation systems and their interrelationships with other aspects of total transportation.

It will concern itself with the planning, design, construction, operation, maintenance and safety of facilities and their components; the economics, financing and administration of the systems and their interactions with the physical, economic, legal and social environment they are designed to serve.

5. *Functions*

The major functions of the Board will be accomplished through the voluntary efforts of those who are associated with activities of the Board, enhanced to the extent feasible by efforts of a staff and to the extent practical under support that may be made available. Funds provided by sponsoring agencies shall be used to support those activities stipulated by the agreements with the sponsors.

a. *Regular Functions* of the Board will be those activities designed to: Identify research needs and resources; stimulate research efforts; provide for the exchange of concepts and techniques relating to research; foster the reporting, evaluation, synthesis and interpretation of research findings, and provide for their publication where appropriate; promote coordination of national research efforts; and encourage the dissemination and utilization of knowledge derived from research.

b. *Special Functions* of the Board will be those approved by its Executive Committee and accepted by the National Research Council as appropriate, for which special financing has been arranged. Special function activities shall be those directed towards or involving consultation, advisory, research or administrative services to governmental or other sponsoring agencies.

6. *Supporting and Sustaining Members*

a. *Supporting Members.* Supporting Members of the Highway Research Board may be any individuals, groups of individuals, firms, corporations, associations, governmental departments or agencies, or institutions or organizations of other kind or nature interested in and desirous of providing financial support for the regular activities of the Board. Application for membership shall be subject to approval by the Executive Committee of the Board. The Committee shall determine and decide the privileges of such Supporting Members and the amount of dues payable by them. Supporting Members shall be classified according to the groups or categories specified and designated as follows:

(1) *Governmental Agencies and Academic Organizations.* This group or category shall extend to and include (a) departments, bureaus, and other subdivisions and units of federal, state and local governments and foreign govern-

ments and their agencies; (b) educational institutions; (c) professional societies; and (d) transportation agencies quasi-governmental in character.

(2) *Private Organizations.* The membership of this group or category shall comprise and extend to business-oriented organizations of whatsoever kind or nature, both domestic and foreign, including but not limited to business and industrial corporations, firms and associations, consultants, business and trade associations, research organizations and service organizations.

(3) *Individual Supporting Members.* There shall be included in this group all persons accepted for membership in an individual capacity.

b. *Sustaining Members.* This group or category shall comprise and include all those qualifying under a(1), (2), or (3) above, whose financial participation equals or exceeds the amount prescribed by the Executive Committee for qualification and acceptance as a Sustaining Member. The Executive Committee shall prescribe the privileges of Sustaining Members.

7. *Executive Committee*

a. *Composition.* The Executive Committee of the Highway Research Board shall consist of not more than twenty appointed members, who shall suitably represent all the major interests of the Board, and not more than eight ex officio members.

b. *Ex Officio Members.* The ex officio members of the Executive Committee, not to exceed eight in number, shall be as follows: (a) the two immediate past Chairmen of the Executive Committee; (2) the Chairman of the Division of Engineering of the National Research Council; (3) the Federal Highway Administrator; (4) the Executive Director of the American Association of State Highway Officials and (5) the chief administrative officers of governmental departments or agencies or other organizations which are or shall become major financial supporters of the regular functions of the Board.

c. *Nomination of Appointive Members.* Each year prior to the annual meeting of the Highway Research Board the Chairman of the Executive Committee shall appoint a nominating committee, collectively representing the scope of the Board's interests, which shall nominate individuals for appointment to the Executive Committee to fill the vacancies occasioned by the expiration of the terms of members. The names of nominees so selected shall be submitted by and through the Executive Director of the Highway Research Board to the Chairman of the Division of Engineering of the National Research Council.

d. *Designation of Ex Officio Members.* Designation of new ex officio members, holding the positions described in Paragraph 7b(5) hereof, shall be by formal action of the Executive Committee and with the approval of the Chairman of the Division of Engineering.

e. *Appointment.* Appointments to membership on the Executive Committee shall be made by the Chairman of the Division of Engineering of the National Research Council, with the approval and consent of the President of the National Academy of Sciences. Such appointments shall be duly made in advance of the annual meeting of the Highway Research Board.

f. *Term of Office.* The term of office of each appointive member of the

Executive Committee shall be for a period of three years. Each appointive member shall be eligible to serve for three consecutively appointed terms of three years each. Upon the expiration of this third consecutive term a member may not be reappointed until one year has elapsed from the expiration of such third consecutive term. Thereafter he may again be appointed for a three-year term and shall be eligible to serve three consecutive terms of three years each. Provided, however, that a Chairman of the Executive Committee upon the completion of his one-year term in that office shall vacate the remaining unexpired term, if any, of his appointive office on the Committee and shall then serve as an ex officio member of the Committee for a two-year period. Upon termination thereof, he shall be eligible for reappointment to a full three-year term and at the close of such period he shall not then be eligible for reappointment until the lapse of one year.

g. *Election of Chairman and Vice Chairmen.* There shall be elected each year in advance of the annual meeting of the Highway Research Board a Chairman, First Vice Chairman, and Second Vice Chairman of the Executive Committee. Election shall be by written ballot cast by the members of the incoming Executive Committee. Nomination for election to such offices shall be made by a nominating committee, appointed from the members of the outgoing Executive Committee by the Chairman of that Committee. Ex officio members of the Executive Committee shall not be eligible for election to the offices of Chairman or Vice Chairmen.

h. *Duties and Term of Office of Chairman and Vice Chairmen.* The Chairman and Vice Chairmen shall be invested with the powers and duties usual and incident to such offices. The terms of office of each the Chairman, First Vice Chairman and Second Vice Chairman shall be for a period of one year, commencing at the close of the regular general session of the annual meeting of the Highway Research Board.

i. *Assumption of Duties by Members of Executive Committee.* The members of the incoming Executive Committee shall assume the duties of their offices upon termination of the regular general session of the annual meeting of the Board, except as otherwise hereinabove provided with respect to the election of a Chairman and Vice Chairmen.

j. *Vacancies.* In the event a vacancy shall occur on the Executive Committee for a reason other than normal expiration of term, such vacancy shall be filled by and through nomination and appointment in the same manner as prescribed in Subsections c and e hereof for the nomination and appointments of members to full three-year terms. A member so appointed shall serve for the remainder of the unexpired term to which he was appointed. He shall then be eligible to serve for two full terms of three years each.

k. *Powers and Duties.* The Executive Committee shall advise and recommend to the National Research Council on the policies and with respect to the administration of the Highway Research Board.

8. Officers

a. *Chairman, Vice Chairmen and Executive Director.* The officers of the Highway Research Board shall be the Chairman, First Vice Chairman and Second Vice Chairman of the Executive Committee and the Executive Director of the

Highway Research Board. They shall be invested with all such powers and duties as are necessary to the supervision and management of the actvities of the Board.

b. *Executive Director.* The Executive Director of the Highway Research Board shall be a full-time employee of the National Academy of Sciences assigned through the Division of Engineering of the National Research Council to the Board upon nomination of the Executive Committee. The Executive Director shall be the chief executive officer of the Board and shall serve as Secretary of the Executive Committee. He shall be responsible for the administration of the Board's activities and shall account to the Executive Committee for all funds received and expended by the Board. The Executive Committee shall make recommendations as to the salary of the Executive Director.

9. Voting

The right to vote with respect to all matters relating to the organization and operations and policies of the Highway Research Board shall be vested solely in the members of the Executive Committee. A quorum shall consist of fifteen members of the Executive Committee. Any action shall be based on a simple majority of the votes cast.

10. Steering Committee

The Steering Committee shall consist of the four officers designated in Paragraph 8a hereof and the ex officio members of the Executive Committee designated in Paragraph 7b hereof.

The Steering Committee may be called by the Executive Director with the approval of the Chairman to advise on actions to be taken within the policies established by the Executive Committee.

11. Divisions and Groups

The functions of the Board shall be carried out under an organizational structure of four Divisions:

Division A—Regular Technical Activities

The organization of Division A will provide three major groups of committees and task forces to consider the concerns of the Board as stated in its scope (Paragraph 4 hereof).

Division B—Special Technical Activities

Division C—Administrative Activities

Division D—National Cooperative Highway Research Program Administration

Changes in this basic organizational structure may be made upon recommendation of the Executive Committee and approval of the Chairman of the Division of Engineering.

The following Senior Committees and Councils shall be established to assist the Executive Committee in its functions with respect to the work of the Board:

Division A Council

Councils for each of the three groups of Division A

Division B Steering Committee

Division C Administrative Committees
Division D Advisory Committee

The Chairmen and members of Senior Committees and Councils will be appointed by the Chairman of the Executive Committee. The terms of office shall be for a period of three years and members shall be eligible for reappointment except that Chairmen shall be limited to two successive terms.

12. Committees and Task Forces

The Senior Committees and Councils shall, subject to the approval of the Executive Committee, establish such subordinate organizations including committees and task forces as are deemed necessary and required for the conduct of the regular functions (described in Paragraph 5a hereof) of the Board. The chairmen and members of such committees and task forces shall be appointed by the Executive Director with the approval of the Chairman of the Executive Committee. The terms of office shall be for a period of three years and members shall be eligible for reappointment, except that the chairmen shall be limited to two successive terms.

13. Special Advisory Committees

Committees shall be established to advise and recommend with respect to the conduct of all special function activities (described in Paragraph 5b hereof) undertaken by the Board. Members of such committees shall be appointed by the Chairman of the Executive Committee, subject to the approval of the Chairman of the Division of Engineering of the National Research Council.

14. Financial and Business Management

The financial and business affairs of the Highway Research Board shall be conducted in conformity with established procedures of the National Academy of Sciences and the National Research Council relating thereto. The National Academy of Sciences, acting through its Council and its Executive Officers, will be responsible for the use of the funds of the Board for purposes of the Board.

15. Meetings

The annual and other meetings of the Highway Research Board shall be held at such times and places as are designated by the Executive Committee. The Executive Committee shall convene during the annual meeting of the Board and other meetings of the Committee may be held at the call of its Chairman upon twenty days' notice in writing to the members. Business at all meetings of the Executive Committee shall be conducted in accordance with Robert's Rules of Order.

16. Amendments to Bylaws

Amendments to these bylaws may be made pursuant to majority vote of the members of the Executive Committee. Provided, however, that before an amendment may become effective, approval thereof by the Governing Board of the National Academy of Sciences and the National Research Council shall be first obtained.

APPENDIX H *Publications Series of the Highway Research Board*

Series Title	Publication Frequency	Designation	Inclusive Dates
Annual Report	Annually		1946 to present
Highway Research Abstracts	Monthly	No. 1 to 147	1931 to June 1947
		Vol. 17, No. 7 . . .	July 1947 to present
Highway Research Circulars	Irregular intervals	No. 1, . . .	1965 to present
Highway Research Correlation Circular		No. 1 to 491	1956 to 1962
Highway Research In Progress[a]	Annually	No. 1, . . .	1968 to present
Highway Research News[b]	Quarterly	No. 1, . . .	1963 to present
Highway Research Record	Irregular intervals	No. 1, . . .	1963 to present
Highway Research Review	Irregular intervals	Series 1, No. 1 to 5	1950 to 1962
HRB Bibliography[c]	Irregular intervals	No. 1, . . .	1947 to present
HRB Bulletin		No. 1 to 362	1946 to 1962
HRB Current Road Problems[d]		No. 1 to 13	1942 to 1946
HRB Proceedings	Annually	Vol. 1 to 41	1921 to 1962
HRB Roadside Development Committee Report			1936 to 1962
HRB Research Report	Irregular intervals	No. 1 to 18C	1945 to 1958
HRB Special Report	Irregular intervals	No. 1, . . .	1952 to present
HRIS Abstracts[e]	Quarterly	Vol. 1, No. 1, . . .	March 1968 to present
NCHRP Report	Irregular intervals	No. 1, . . .	1964 to present
NCHRP Synthesis of Highway Practice	Irregular intervals	No. 1, . . .	1969 and continuing
Yearbook	Annually	No. 1, . . .	1946; 1962 to 1967

[a] Developmental issues published September 1965 and April 1967.
[b] Earlier editions published from October 1925 to March 1927.
[c] Seven nonserials published in 1965.
[d] No. 13 revised in 1957 (No. 13-2R).
[e] Experimental issues published October 1967 and December 1967.

APPENDIX I *Organization of Departments and Departmental Committees*

DEPARTMENTS*

A. *List of Departments*

Department of Economics, Finance and Administration
Department of Design
Department of Materials and Construction
Department of Maintenance
Department of Traffic and Operations
Department of Soils, Geology and Foundations
Department of Urban Transportation Planning
Department of Legal Studies

B. *Purpose*

The purpose of each department is to plan an overall research program and suggest research projects; to encourage, correlate and evaluate research; to assist in outlining sound procedures for conducting such research; to encourage the reporting and discussion of research findings before the open forum of the Board's meetings; to develop plans for effective dissemination and utilization of research findings; to suggest to the Board any action it believes should be taken for the good of the Board; and to provide the Board with an annual report on departmental and committee activities. The scope of each department is as indicated by its title.

C. *Functions*

1. Maintain a comprehensive plan of needed research within the scope of the department.
2. Review research activities of departmental committees to see how well these needs are being meet.
3. Make a periodic review of the department organization, including scopes of divisions and committees, and recommend such changes as may be desirable.
4. Promote the conduct of needed research and stimulate the presentation of research reports.
5. Review committee reports and make recommendations concerning action on them.
6. Review papers not falling within the scope of the departmental committees and recommended disposition of same.
7. Consider specific problems submitted to it by the Executive Director and make recommendations.
8. Perform other appropriate functions as desirable.

TECHNICAL COMMITTEES

A. *Purpose*

The basic purposes of all technical committees are essentially the same

* Superseded by Groups in 1970.

as those of the departments, but are more limited in scope and have more specific direction and interest.

B. *Scopes*

Each committee operates within a scope approved by the Executive Committee of the Highway Research Board. Changes in the name or scope of a committee can be recommended by the department chairman to the Executive Director. No change is effective until announced by the Executive Director.

C. *Objectives*

Each committee should adopt a statement of objectives that should delineate immediate courses of action or attack on the problems of highest priority. This statement should be reviewed annually and revised as needed.

D. *Functions*

1. Suggest and plan research.
2. Correlate and evaluate research results.
3. Stimulate the presentation of research reports.
4. Review papers and other documents and recommend the disposition of them.
5. Maintain a list and outline of needed research within the scope of the committee.
6. Make a periodic review of the committee's scope and recommend such changes as may be desirable.
7. Perform other functions as desirable.

APPENDIX J *Evolution of Technological Structure*

DIVISION A—REGULAR TECHNICAL ACTIVITIES

Group 1—Transportation System Planning and Administration
J. Douglas Carroll, Chairman, Group 1 Council

Group 1 was formed in January 1970 from the following predecessor organizations.

1919-1944: *Committee on Economic Theory of Highway Improvements*
Thomas R. Agg, Chairman, 1919-1929
Roger L. Morrison, Chairman, 1932-1944
Renamed as the Committee on Highway Transportation Costs in 1929, it became the Department of Highway Transportation Economics in 1935 and continued with this name until 1944.

1923-1935: *Committee on Highway Finance*
H. G. McKay, Chairman, May 1923
H. R. Trumbower, Chairman, November 1923
Harry J. Kirk, Chairman, 1929

1934-1943: *Department of Finance and Administration*
Thomas H. MacDonald, Chairman, 1935-1943
These two departments were merged in 1944 to become:

1944-1970: *Department of Economics, Finance and Administration*
H. S. Fairbank, Chairman, 1944-1956
Guilford P. St. Clair, Chairman, 1957-1967
Robert C. Blensly, Chairman, 1967-1970

1954-1970: *Department of Urban Transportation Planning*
National Research Council ad hoc Committee on Urbanization Research, 1954-1955
Special Committee on Urban Research
E. Willard Dennis, Chairman, 1955-1961
Pyke Johnson, Chairman, 1961-1969
Reorganized as the Department of Urban Transportation Planning in 1964

1961-1970: *Department of Legal Studies*
Functioned as a Special Committee on Highway Laws in the Department of Economics, Finance and Administration, 1953-1960.
J. H. Beuscher, Chairman, Special Committee on Highway Laws, 1961-1964; Department of Legal Studies, 1964-1967
John P. Holloway, Chairman, Department of Legal Studies, 1967-1970

Group 2—Design and Construction of Transportation Facilities
John L. Beaton, Chairman, Group 2 Council

Group 2 was formed in January 1970 from the following predecessor organizations.

1922-1970: *Department of Design*
This Department began as the Committee on Structural Design of Roads in 1922 and became the Department of Design in 1936.
Albert T. Goldbeck, Chairman, 1922-1935
Carlton N. Conner, Chairman, 1936-1954
Tilton E. Shelburne, Chairman, 1954-1964
W. B. Drake, Chairman, 1964-1970.

1922-1970: *Department of Materials and Construction*
This Department began as the Committee on Character and Use of Road Material in 1922; it was called the Committee on Materials and Construction from 1930-1934.
H. S. Mattimore, Chairman, 1922-1934
Charles H. Scholer, Chairman, 1935-1953
R. R. Litehiser, Chairman, 1954-1961
John H. Swanberg, Chairman, 1962-1964
R. L. Peyton, Chairman, 1964-1970

1935-1970: *Department of Soils, Geology and Foundations*
Chester A. Hogentogler, Chairman, Department of Soils Investigation, 1935-1947

Harold Allen, Chairman, Department of Soils Investigation, 1948-1954

Frank R. Olmstead, Chairman, Department of Soils, 1954; Department of Soils Investigation, 1955; Department of Soils, Geology and Foundations, 1956-1957

Miles S. Kersten, Chairman, Department of Soils, Geology and Foundations, 1958-1963

Eldon J. Yoder, Chairman, Department of Soils, Geology and Foundations, 1964-1970

Group 3—Operation and Maintenance of Transportation Facilities
Harold L. Michael, Chairman, Group 3 Council

Group 3 was formed in January 1970 from the following predecessor organizations.

1921-1970: *Department of Maintenance*

W. H. Root, Chairman, Committee on Maintenance, 1921-1928
G. C. Dillman, Chairman, Committee on Maintenance, 1928-1929
B. C. Tiney, Chairman, Committee on Maintenance, 1929-1934
C. P. Owens, Chairman, Department of Maintenance, 1934-1936
C. W. McClain, Chairman, Department of Maintenance, 1936-1937
W. H. Root, Chairman, Department of Maintenance, 1937-1950
Rex M. Whitton, Chairman, Department of Maintenance, 1950-1952
V. L. Ostrander, Chairman, Department of Maintenance, 1952-1956
M. W. Fisher, Chairman, Department of Maintenance, 1956-1957
H. E. Diers, Chairman, Department of Maintenance, 1957-1963
H. J. Rathfoot, Chairman, Department of Maintenance, 1963-1967
John P. Murphy, Chairman, Department of Maintenance, 1967-1968
J. F. Andrews, Chairman, Department of Maintenance, 1968-1970

1922-1970: *Department of Traffic and Operations*

George E. Hamlin, Chairman, Committee on Highway Traffic Analysis, 1922-1930; Committee on Traffic, 1930-1932
William A. Van Duzer, Chairman, Committee on Traffic, 1932-1935
Charles J. Tilden, Chairman, Department of Traffic, 1935-1938
Arnold H. Vey, Chairman, Department of Traffic, 1938-1942
Wilbur S. Smith, Chairman, Department of Traffic and Operations, 1942-1957
Donald S. Berry, Chairman, Department of Traffic and Operations, 1957-1960
Fred W. Hurd, Chairman, Department of Traffic and Operations, 1960-1964
Harold L. Michael, Chairman, Department of Traffic and Operations, 1964-1970

APPENDIX K *Awards Presented by the Highway Research Board*

THE ROY W. CRUM DISTINGUISHED SERVICE AWARD

The Highway Research Board in 1948 established an award to be made in recognition of outstanding achievement in the field of highway research. This award was known as the "Highway Research Board Distinguished Service Award" until 1952 when the Executive Committee redesignated it the ROY W. CRUM DISTINGUISHED SERVICE AWARD as a memorial to the Board's late director, Roy W. Crum, who had served as head of the HRB staff from 1928 until his death in 1951.

Outstanding achievement consists of distinguished service, production of fundamental or developmental research, or the administration, promotion, or fostering of outstanding research which in the judgment of the Executive Committee is worthy of the Award. The following men have received the Award:

Charles H. Scholer, 1948
Frank H. Jackson, 1948
Kenneth B. Woods, 1949
O. K. Normann, 1949
Fred V. Reagel, 1950
Albert T. Goldbeck, 1950
Roy W. Crum (posthumously), 1951
Herbert S. Fairbank, 1952
Prevost Hubbard, 1953
Charles R. Waters, 1953
Burton W. Marsh, 1954
Ralph A. Moyer, 1954
Walter H. Root (posthumously), 1954
Earl Foster Kelley, 1955
Tilton E. Shelburne, 1955
Stanton Walker, 1955
Francis N. Hveem, 1956
Edward H. Holmes, 1957
Harmer E. Davis, 1958
Guilford P. St. Clair, 1959
Merlin G. Spangler, 1959
Robert R. Litehiser, 1960
Harold Allen, 1960
Elmer M. Ward (posthumously), 1961
Rex M. Whitton, 1961
William H. Goetz, 1961
Alvin C. Benkelman, 1962
David R. Levin, 1962
Fred Burggraf, 1963
Carl C. Saal, 1964
Kurt Frank Wendt, 1964
W. S. Housel, 1965
Bryant Mather, 1966
Robert F. Baker, 1967
John H. Swanberg, 1968
Francis C. Turner, 1969

THE HIGHWAY RESEARCH BOARD AWARD

The HIGHWAY RESEARCH BOARD AWARD was instituted in 1940 to give recognition to the authors of papers of outstanding merit presented at the Annual Meetings of the Highway Research Board. The following men have received the HIGHWAY RESEARCH BOARD AWARD:

Wilfred Owen, 1940
F. V. Reagel, 1941
M. B. Russell, 1942
M. G. Spangler, 1942
Ralph A. Moyer, 1943
Donald J. Belcher, 1944
Donald W. Loutzenheiser, 1945
K. B. Woods, 1946

Harold S. Sweet, 1946
T. E. Shelburne, 1946
Norman W. McLeod, 1947
C. Raymond Hanes, 1948
F. N. Hveem, 1949
R. M. Carmany, 1949
Roy E. Jorgensen, 1950
Robert G. Mitchell, 1950
Thomas J. Carmichael, 1951
Charles E. Haley, 1951
Hugo C. Duzan, 1952
William R. McCallum, 1952
Thomas R. Todd, 1952
Earl C. Sutherland, 1953
Harry D. Cashell, 1953
C. A. Rothrock, 1954
Carl C. Saal, 1955
Chester McDowell, 1956
George M. Webb, 1957
Karl Moskowitz, 1957
Alan M. Voorhees, 1958
Charles J. Keese, 1959
Charles Pinnell, 1959
William R. McCasland, 1959
Kenneth A. Stonex, 1960
A. S. Lang, 1961
David H. Robbins, 1961
George Haikalis, 1961
Hyman Joseph, 1961
Leonard Casciato, 1962
Sam Cass, 1962
W. Ronald Hudson, 1963
Mark Morris, 1963
David W. Hadley, 1964
Marvin D. Oosterbaan, 1965
G. A. Leonards, 1965
Conrad L. Dudek, 1966
C. J. Keese, 1966
Donald R. Drew, 1966
J. A. Wattleworth, 1967
Frank Gerig, Jr., 1967
Malcolm D. Graham, 1967
Robert H. Freer, 1967
C. H. Oglesby, 1968
Matthew J. Altenhofen, 1968
G. R. Cudney, 1968
Aleksandar S. Vesic, 1969
Surendra K. Saxena, 1969
Jack E. Leisch, 1969

THE FRED BURGGRAF AWARD

The FRED BURGGRAF AWARD was established in 1966 to stimulate and encourage young researchers to contribute to the advancement of knowledge in the field of highway transportation. It provides recognition of excellence in highway research by researchers 30 years of age or younger who have presented papers at the Board's Annual Meetings. The Award carries with it a cash prize and was named in honor of the late Fred Burggraf, who served as the Board's Director from 1951 until his retirement in 1964. The following men have received the FRED BURGGRAF AWARD:

John E. Taylor, 1967
Ronald W. Tweedie, 1967
Thomas G. Williamson, 1967
Matthew W. Witczak, 1967
R. D. Worrall, 1968
Ira F. Doom, 1968
Thomas G. Williamson, 1969
Peter R. Stopher, 1969

THE GEORGE S. BARTLETT AWARD

The Award is named in honor of the late George S. Bartlett, a pioneer in highway development and technology. The Award was established to recognize and be conferred annually upon some individual who has made an outstanding contribution to highway progress. It was first presented in 1931. The instrument which created the endowment for the Award assigned the selection of the recipient to a three-member Board of Award made up of representatives from the American Association of State Highway Officials, the American Road Builders' Association, and the Highway Research Board.

The three sponsoring organizations rotate the chairmanship of the Board of Award, as well as the occasions at which the Award is presented. The following men have received the GEORGE S. BARTLETT AWARD:

Thomas H. MacDonald, 1931
Arthur N. Johnson, 1932
James H. MacDonald, 1933
Frank F. Rogers, 1934
Edward N. Hines, 1935
Thomas R. Agg, 1936
C. A. Hogentogler, 1937
Fred R. White, 1938
Robert Moses, 1939
Frank T. Sheets, 1941
Paul G. Hoffman, 1942
H. S. Mattimore, 1943
Charles H. Purcell, 1944
Frederick E. Everett, 1945
Charles M. Upham, 1946
H. S. Fairbank, 1947
G. Donald Kennedy, 1948
M. J. Hoffman, 1949
R. H. Baldock, 1950
C. S. Mullen, 1951
Roy W. Crum (posthumously), 1951
Samuel C. Hadden, 1952
D. C. Greer, 1953
James A. Anderson, 1954
William Randolph Hearst, Jr., 1955
Pyke Johnson, 1956
C. D. Curtiss, 1957
Rex M. Whitton, 1958
George H. Fallon, 1959
W. A. Bugge, 1960
Bertram D. Tallamy, 1961
A. E. Johnson, 1962
Julian R. Steelman, 1963
Francis C. Turner, 1964
Louis W. Prentiss, 1965
J. C. Womack, 1966
J. B. McMorran, 1967
David H. Stevens, 1968
Harmer E. Davis, 1969

APPENDIX L *Research Projects Undertaken by Highway Research Board Before Formation of Special Projects Division and NCHRP*

More than 40 research projects and investigations were administered by the Highway Research Board between 1920 and 1963. Since then the Special Projects Division and NCHRP have administered many more, which are listed in Appendixes N and O. Among those administered 1920-1960 are the following:

DATE	TITLE
1924-25	Economic Value of Steel Reinforcement in Concrete Pavements
1925-28	Earth Road Investigation (Low-Cost Improved Roads)

DATE	TITLE
1925-32	Culvert Investigation
1928-33	Curing Investigation (Concrete Pavements)
1929-32	Rail Steel Investigation (for use in highway construction)
1930-32	Traffic Survey Methods and Forms
1931	Compaction of Fills
1931	Tractive Resistance and Allied Problems
1931-32	Mineral Aggregate Research
1931	Fillers and Cushion Courses for Brick and Black Pavement
1931-32	Dust Laying Investigation (Use of $CaCl_2$)
1932	Study of the Laws, Funds, Organizations and Technical Practices Relating to Roadside Development (joint with AASHO)
1933-39	High Elastic Steel as Concrete Reinforcement
1933-41	Investigation of Warping Concrete Slabs
1933-34	Evaluation of Highway Systems (Improvements)
1934-36	Stabilized Roads Investigation
1935-52	Development of Equipment for Stabilized Roads (joint with ARBA)
1936-43	Safety Research Project (from Department of Agriculture)
1940-42	Durability of Concrete (joint ASTM, ACI, HRB)
1942	Traffic Control Devices (joint with Eno Foundation)
1942	Bearing Power of Timber Piles
1943-45	Freeze-Proofing of Soil With Calcium Chloride
1943-55	Hybla Valley Non-Rigid Pavement Project (joint with BPR and AI)
1946-63	Calcium Chloride Stabilization
1948-49	Motor Vehicle Size-Weight Study (Pennsylvania Turnpike)
1949-58	Vehicle Characteristics
1950-53	Road Test One–MD
1951-53	Study of Intergovernmental Relationship
1951-54	Study of Wind Stresses and Bridges (Cooperative Project)
1951-56	Economics of Parking
1951-57	Maryland Intergovernmental Relationship
1952-54	WASHO Road Test
1954-56	Effects of Military Aircraft
1954-58	Urbanization Research
1954-61	Highway Laws Study
1955-62	AASHO Road Test
1957-62	Friction Piles for Highway Structures
1957-58	Highway Finance and Taxation (joint with BPR and ASF)
1959-62	Framework Study for Urban Research
1961-63	Highway-Public Utility Liaison
1961-63	Motor Vehicle Registration and Titling Practices

APPENDIX M *Outlines of Highway Research Needs Drafted by the Highway Research Board, 1920-1966*

The Highway Research Board has drafted proposals for nationwide research activities on highway needs on several occasions. A distinguished predecessor of these statements was Anson Marston's paper presented at the AASHO Annual Meeting in December 1919 (see Chapter 6). The following proposals have been published by the Board in the ensuing 50 years.

August 1921—A "tentative draft" was prepared for limited distribution. No copies exist, but it is assumed that the "Suggested Outline of Activities in Highway Research" incorporated as an appendix (pages 28-30) of the minutes of the First Annual Meeting may be the listing drafted in August 1921. The published statement covers 24 problems distributed among 8 broad areas for committee activity.

December 1929—A "Proposed Program of Highway Research" is noted on pages 24-51 of the *Proceedings of the Ninth Annual Meeting.* A discussion of needs and listing of research problems included 6 fields and 130 problems. The breakdown is as follows: Administration and Finance, 12 problems; Design, 52; Transportation Costs, 8; Materials and Construction, 21; Maintenance, 9; and Traffic, 28.

July 1946—A 67-page "National Research Program" was printed in mimeograph form and referred to in the minutes of the Annual Meeting of the Highway Research Board, December 6, 1946 (pages 614-615). Some 810 separate items were listed among the six departments of the Board. This listing was obtained by asking state highway departments to list problems for further research.

June 1959—*Highway Research in the United States: Needs, Expenditures and Applications, 1959* was published as Highway Research Board Special Report 55. Its 119 pages included an Appendix A, "Highway Research in the United States: An Analysis of Current Fiscal Support"; an Appendix B, "Highway Research Problems of Importance," with 101 problems listed; and an Appendix C, "Applications of Highway Research Problems: Examples Reported by State Highway Departments."

1965—A canvass of needs was made by departments and committees of HRB for the proposed development of a framework for a nationwide research needs program. Some 1,000 problem statements were assembled that provided source information for the development of a Highway Research Needs Program for NCHRP.

Research Needs Statements Prepared by HRB Departments

ECONOMICS, FINANCE AND ADMINISTRATION

Highway Research Organizations: Descriptions of Existing Organizational Patterns and Scope of Activities, by M. Earl Campbell, was published as Highway Research Board Special Report 15 in 1953.

"A Research Census in Economics, Finance and Administration Related to the Highway Field," a 22-page record obtained by mail canvass, was published in mimeograph form in December 1959.

DESIGN

The Committee on Composite Pavement Design prepared "Suggestions for the Study of Composite Pavement," published as Highway Research Correlation Service Circular 473 in July 1962.

The Committee on Flexible Pavement Design compiled "A Guide for Experimental Flexible Pavement Projects," published as Highway Research Correlation Service Circular 399 in September 1959.

The Committee on Photogrammetry and Aerial Surveys produced "Research Needed in Photogrammetry and Aerial Surveys for Highways," published as Highway Research Correlation Service Memo No. 4, Circular 480, September 1962. The committee also prepared "Research in Photogrammetry and Aerial Surveys for Highways: A Supplemental Report on Five Problem Statements" as a mimeographed Memo Report in September 1964.

The Committee on Rigid Pavement Design compiled "Recommendations for Experimental Continuously Reinforced Concrete Pavements." It was published as Highway Research Correlation Service Circular 372 in November 1958.

The Committee on Surface Drainage of Highways published "Needed Research in Highway Drainage," Highway Research Correlation Service Memo No. 1, Circular 462, in February 1962.

Research Needed in Geometric Highway Design, by D. W. Loutzenheiser, was published as Highway Research Board Special Report 12, 1953.

MATERIALS AND CONSTRUCTION

The Ad Hoc Committee on Research Problems of Mutual Interest and Concern to Users and Producers of Asphaltic Materials prepared "Characteristics of Asphaltic Materials and Mixing and Placing of Asphalt Concrete Mixtures." It was published as Highway Research Correlation Service Circular 452 in October 1961.

APPENDIX N *National Cooperative Highway Research Program Projects for Fiscal Years 1963 Through 1971*

PROJECT NO.	TITLE	RESEARCH AGENCY	CONTRACT AMOUNT OR CONTRACT COST
	Area One: Design—Pavements		
1-1(1)	Development of Procedures for Comparing the AASHO Road Test Findings with Performance of (1) Existing Pavements and (2) Newly Constructed Experimental Pavements	HRB	$ 42,800*

PROJECT NO.	TITLE	RESEARCH AGENCY	CONTRACT AMOUNT OR CONTRACT COST
1-1(2)	Guidelines for Extending the Findings of the AASHO Road Test—Implementation Phase	HRB	11,356*
1-2	Comparison of Different Methods for Evaluating Pavement Conditions	Purdue U	29,957*
1-3(1)	Factors Influencing Pavement Performance—Regional	Purdue U	46,000
1-3(2)	Factors Influencing Pavement Performance—Local	Northwestern U	19,850*
1-3(3)	Factors Influencing Pavement Performance	U of California	19,800*
1-4(1)	Extension of Road Test Performance Concepts	Georgia Tech	10,000*
1-4(1)A	Extension of Road Test Performance Concepts	Duke U	19,924*
1-4(2)	Extension of Road Test Performance Concepts	Purdue U	12,243*
1-5	Detecting Variations in Load-Carrying Capacity of Flexible Pavements	Cornell Aero Lab	49,011*
1-5(2)	Detecting Seasonal Changes in Load-Carrying Capabilities of Flexible Pavements	Texas A & M	48,050
1-6	Standard Measurements for Satellite Program—Measurement Team	Texas A & M	61,353*
1-7	Development of Interim Skid-Resistance Requirements for Highway Pavement Surfaces	Penn State U	24,815*
1-8	Factors Involved in the Design of Asphalt Pavement Surfaces	Materials R & D	24,869
1-9	Evaluation of Studded Tires	Cornell Aero Lab	24,998*
1-10	Translating AASHO Road Test Findings—Basic Properties of Pavement Components	Materials R & D	195,448
1-11	Evaluation of AASHO Interim Guides for Design of Pavement Structures	Materials R & D	63,720
1-12	Determination of Pavement Friction Coefficients Required for Driving Tasks	Franklin Inst	299,990
1-12(2)	Locked-Wheel Pavement Skid Tester Correlation and Calibration Techniques	—	Contract Pending
	Area Two: Administration—Economics		
2-1	Criteria for Highway Benefit Analysis	U of Washington	101,948*
2-2	Guidelines for the Determination of Community Consequences	U of Washington	48,873*
2-3	Analysis of Motor Vehicle Accident Data as Related to Highway Classes and Design Elements	Cornell Aero Lab	155,972*
2-4	The Value of Highway Travel Time, Comfort, Convenience, and Uniform Driving Speed	Texas A & M	77,100
2-5	Running Cost of Motor Vehicles as Affected by Highway Design and Traffic	Catholic U	101,263*
2-5A	Running Cost of Motor Vehicles as Affected by Highway Design and Traffic	Paul J. Claffey	68,719

PROJECT NO.	TITLE	RESEARCH AGENCY	CONTRACT AMOUNT OR CONTRACT COST
2-6	Warranted Levels of Improvement for Local Rural Roads	Stanford U	40,000*
2-7	Road User Costs in Urban Areas	Catholic U	99,376*
2-8	Estimation and Evaluation of Diverted and Generated (Induced) Traffic	Northwestern U	40,000*
2-9	Effect of Highway Landscape Development on Nearby Property	Franklin Inst	149,103*
2-10	Future Needs for Oversize-Overweight Permit Operation on State Highways	Jorgensen & Assoc	99,655*
2-11	Summary and Evaluation of Economic Consequences of Highway Improvments	HRB	100,000
	Area Three: Traffic—Operations and Control		
3-1	Development of Criteria for Evaluating Traffic Operations	Cornell Aero Lab	158,878*
3-2	Surveillance Methods and Ways and Means of Communicating with Drivers	Cornell Aero Lab	246,756*
3-3	Sensing and Communication Between Vehicles	Ohio State U	163,190*
3-4	Means of Locating Disabled or Stopped Vehicles and Methods of Communication with a Central Location	Airborne Instr	127,991*
3-5	Improved Criteria for Designing and Timing Traffic Signal Systems	Planning Res	264,902*
3-6	Effect of Regulatory Devices on Intersectional Capacity and Operation	De Leuw, Cather	153,175*
3-7	Establishment of Standards for Highway Noise Levels	Bolt Beranek	216,840
3-8	Factors Influencing Safety at Highway-Rail Grade Crossings	Voorhees & Assoc	91,421
3-9	Analysis and Projection of Research on Traffic Surveillance, Communication, and Control	Jorgensen & Assoc	23,760*
3-10	Application of Vehicle Operating Characteristics to Geometric Design and Traffic Operations	Cornell Aero Lab	41,520*
3-11	Optimizing Street Operations Through Traffic Regulations and Control	Peat, Marwick	229,090
3-12	Development of Information Requirements and Transmission Techniques for Highway Users	Airborne Instr	300,500
3-13	Guidelines for Medial and Marginal Access Control of Major Roadways	Texas A & M	149,916
3-14	Optimizing Flow on Existing Street Networks	Edwards & Kelcey	990,000
3-15	Weaving Area Operations Study	Poly of Brooklyn	100,000
3-16	Freeway Lane Drops	System Dev Corp	99,789
3-17	Improving Traffic Operations and Safety at Exit Gore Areas	—	Contract Pending

PROJECT NO.	TITLE	RESEARCH AGENCY	CONTRACT AMOUNT OR CONTRACT COST
	Area Four: Materials and Construction—General Materials		
4-1	Development of Appropriate Methods for Evaluating the Effectiveness of Stabilizing Agents	U of Illinois	114,991*
4-2	A Study of Degrading Aggregates in Bases and Subbases with Production of Excessive Amounts of and/or Harmful Types of Fines	Purdue U	63,990*
4-3(1)	Development of Methods to Identify Aggregate Particles Which Undergo Destructive Volume Changes When Frozen in Concrete	V P I	43,337*
4-3(2)	Development of Methods to Identify Aggregate Particles Which Undergo Destructive Volume Changes When Frozen in Concrete	Penn State U	106,213*
4-4	Synthetic Aggregates for Highway Uses	Battelle Mem Inst	14,790*
4-5	A Study of the Mechanism Whereby the Strength of Bases and Subbases is Affected by Frost and Moisture	Michigan Tech U	64,105*
4-6	Protective Coatings for Highway Structural Steel	Steel Str Paint	25,000*
4-7	Fatigue Strength of High-Yield Reinforcing Bars	P C A	100,000
4-8	Research Needs Relating to Performance of Aggregates in Highway Construction	V P I	59,493
4-9	Evaluation of Preformed Elastomeric Pavement Joint Sealing Systems and Practices	Utah St Hy Dept	93,494
4-10	Promising Replacements for Conventional Aggregates for Highway Use	U of Illinois	50,000
	Area Five: Traffic—Illumination and Visibility		
5-2(1)	Effects of Illumination on Operating Characteristics of Freeways—Traffic Flow, Driver Behavior, and Accidents	Yale University	145,849*
5-2(2)	Effects of Illumination on Operating Characteristics of Freeways—Driver Response, Visibility, and Visual Discomfort	Ohio State U	81,187*
5-2(3)	Effects of Illumination on Operating Characteristics of Freeways—Driver Discomfort	Inst for Research	37,460*
5-3	Visual Information Needed by the Driver at Night	Ohio State U	100,940*
5-4	Economic Study of Roadway Lighting	Franklin Inst	19,412*
5-5	Nighttime Use of Highway Pavement Delineation Materials	Sw Research Inst	150,000
5-6	Highway Fog	Cornell Aero Lab	99,955
5-6A	Highway Fog	—	Contract Pending
5-7	Roadway Delineation Systems	Penn State U	471,250
5-8	Warrants for Highway Lighting	Texas A & M	199,627

PROJECT NO.	TITLE	RESEARCH AGENCY	CONTRACT AMOUNT OR CONTRACT COST
	Area Six: Maintenance—Snow and Ice Control		
6-1	Development of Economical and Effective Chemical Deicing Agents to Minimize Injury to Highway Structures and Vehicles	IIT Research Inst	40,000*
6-2	Nonchemical Methods for Preventing or Removing Snow and Ice Accumulations on Highway Structures	Jorgensen & Assoc	25,000*
6-3	Development and Evaluation of Protective Coatings to Prevent Deterioration of Concrete Structures by Deicing Agents	Battelle Mem Inst	58,557*
6-4	Evaluation and Development of Methods for Reducing Corrosion of Reinforcing Steel	Battelle Mem Inst	39,330*
6-5	Study of Physical Factors Influencing Resistance of Concrete to Deicing Agents	U of Illinois	72,500*
6-6	To Evaluate Existing Methods and/or Develop Improved Methods for the Measurement of Certain Properties of Concrete	Ohio State U	69,393*
6-7	Estimation of Disintegration in Concrete Structures	Geotechnics	8,547*
6-7A	Estimation of Disintegration in Concrete Structures	IIT Research Inst	44,614*
6-8	Evaluation of Methods of Replacement of Deteriorated Concrete in Structures	Tallamy Assoc	25,000*
6-9	Potential Accelerating Effects of Chemical Deicing Damage by Traffic and Other Environmental-Induced Stresses in Concrete Bridge Decks	U of Illinois	200,000*
6-10	Develop Improved Snow Removal and Ice Control Techniques at Interchanges	Tallamy Assoc	95,000
6-11	Economic Evaluation of the Effects of Ice and Frost on Bridge Decks	—	Contract Pending
	Area Seven: Transportation Planning—Traffic Planning		
7-1	The Influence of Land Use on Urban Travel Patterns	Louis E. Keefer	129,568*
7-2	Traffic Attraction of Rural Outdoor Recreational Areas	IIT Research Inst	49,496*
7-3	Weighing Vehicles in Motion	Franklin Inst	73,391*
7-4	Factors and Trends in Trip Lengths	Voorhees & Assoc	150,657
7-5	Predicted Traffic Usage of a Major Highway Facility Versus Actual Usage	Yale University	99,675*
7-6	Multiple Use of Lands Within Highway Rights-of-Way	Barton-Aschman	24,220*
7-7	Motorists' Needs and Services on Interstate Highways	Airborne Instr	99,267*
7-8	User Cost and Related Consequences of Alternative Levels of Highway Service	—	Contract Pending

PROJECT NO.	TITLE	RESEARCH AGENCY	CONTRACT AMOUNT OR CONTRACT COST
	Area Eight: Transportation Planning—Urban Transportation		
8-1	Social and Economic Factors Affecting Travel	Vogt, Ivers	94,558*
8-2	Factors Influencing Modal Trip Assignment	IIT Research Inst	298,033*
8-3	Individual Preferences for Various Means of Transportation	U of Penn	63,282*
8-4	Criteria for Evaluating Alternative Transportation Plans	Northwestern U	89,900*
8-4A	Criteria for Evaluating Alternative Transportation Plans	U of Illinois	5,000
8-5	Transportation Aspects of Land-Use Controls	Victor Gruen	125,967*
8-6	Individual Preferences for Alternative Dwelling Types and Environments	U of N Carolina	99,897*
8-7	Evaluation of Data Requirements and Collection Techniques for Transportation Planning	Creighton, Hamburg	190,000
8-8(1)	The Impact of Highways Upon Environmental Values (Study Design)	M I T	30,000
8-8(2)	The Impact of Highways Upon Environmental Values (Study Design)	Daniel, Mann et al	28,950*
8-8(3)	The Impact of Highways Upon Environmental Values	M I T	220,000
	Area Nine: Materials and Construction—Bituminous Materials		
9-1	Asphalt Durability and Its Relation to Pavement Performance	American Oil	100,000*
9-2	Asphalt Durability and Its Relation to Pavement Performance—Adhesion	Montana College	107,670
9-3	Evaluation of Pavement Joint and Crack Sealing Materials and Practices	Rensselaer	24,996*
	Area Ten: Materials and Construction—Specifications, Procedures, and Practices		
10-1	Development of Guidelines for Practical and Realistic Construction Specifications	Miller-Warden	25,000*
10-2	Evaluation of Construction Control Procedures	Miller-Warden	59,750*
10-2A	Evaluation of Construction Control Procedures	Materials R & D	70,945*
10-3	Effects of Different Methods of Stockpiling and Handling Aggregates	Miller-Warden	55,000*
10-4	Rapid Test Methods for Field Control of Construction	Clemson U	99,320*
10-5	Density and Moisture Content Measurements by Nuclear Methods	Res Triangle Inst	88,636*
10-5A	Optimization of Nuclear Density and Moisture Content Measurement Methods	N Carolina State U	49,986

PROJECT NO.	TITLE	RESEARCH AGENCY	CONTRACT AMOUNT OR CONTRACT COST
10-6	Measurement of Pavement Thicknesses by Rapid and Nondestructive Methods	IIT Research Inst	108,821*
10-7	Potential Uses of Sonic and Ultrasonic Devices in Highway Construction	Ohio State U	24,310*
10-8	Evaluating Procedures for Determining Concrete Pavement Thickness and Reinforcement Position	Pa Dept of Hwys	149,835
10-9	Criteria for Need of Seal Coats for Bituminous Pavements	U of Minnesota	36,814
	Area Eleven: Administration—Law		
11-1	Rules of Compensability and Valuation in Highway Land Acquisition	U of Wisconsin	84,840*
11-1(1)	Eliminating Enhancement or Diminution Effects on Right-of-Way Valuation	Real Estate Res	5,000
11-1(2)	Recognition of Benefits to Remainder Property in Highway Valuation	Montano & Assoc	5,000*
11-1(3)	Taxation Aspects of Right-of-Way Acquisition	U of Tulsa	2,500
11-1(4)	Compensation in the Nature of Additives to Market Value	U of Oklahoma	2,500*
11-1(5)	Rules of Discovery and Disclosure in Highway Condemnation Proceedings	Long, Mikkelborg	2,500*
11-1(6)	Valuation and Condemnation Problems of Selected Special Purpose Properties	Edward E. Level	7,500
11-1(7)	Valuation and Compensability of Noise, Pollution, and Other Environmental Factors	U of Oklahoma	2,500
11-1(8)	Remainder Damages Caused by Drainage, Runoff, Blasting, and Slides	Harrison Lewis	7,500
11-1(9)	Valuation and Condemnation Problems Involving Trade Fixtures	Edward L Snitzer	5,000
11-1(10)	Compensability and Valuation Aspects of Residential Displacement in Highway Programs	Ross Hardies et el	5,000
11-1(11)	Valuation Elements of Joint Development Projects, Including Air Rights	Real Estate Res	5,000
11-2	Theory and Practice in Inverse Condemnation	Reg & Urban Plan	15,000*
11-3	Valuation and Legal Implications of Scenic, Conservation, and Roadside Easement	Sutte, Jr. & Assoc	25,000*
11-3(1)	Public Control of Roadside Advertising Signs for Highway Beautification	Sutte, Jr. & Assoc	20,000
11-3(2)	Public Control of Junkyards for Highway Beautification	Real Estate Res	13,300
11-4	Elimination of Wide Divergence in Right-of-Way Valuation	Am Inst Rl Est App	25,000
11-5	Valuation of Air Space	—	Contract Pending
	Area Twelve: Design—Bridges		
12-1	Deformation of Steel Beams Related to Permitted Highway Bridge Overloads	U of Missouri	50,000

PROJECT NO.	TITLE	RESEARCH AGENCY	CONTRACT AMOUNT OR CONTRACT COST
12-2	Distribution of Wheel Loads on Highway Bridges	Iowa State U	79,512*
12-3	Development of Waterproof Roadway Joints for Bridges	Sw Research Inst	149,895
12-4	Thermal Characteristics of Highway Bridges	Sw Research Inst	102,400*
12-5	Protection of Steel in Prestressed Concrete Bridges	U of Denver	173,255
12-6	Prediction of Permanent Camber of Bridges	U of Missouri	85,000
12-7	Effects of Weldments on Fatigue Strength of Steel Beams	Lehigh University	400,000
12-8	Bridge Rail Service Requirements as a Basis for Design Criteria	Texas A & M	100,000
12-9	Elastomeric Bearing Research	Battelle Mem Inst	84,800
12-10	Analysis and Design of Bridge Bents	PCA	250,000
12-11	Waterproof Membranes for Protection of Concrete Bridge Decks	—	Contract Pending
12-12	Welded Steel Bridge Members Under Variable-Cycle Fatigue Loadings	—	Contract Pending
	Area Thirteen: Maintenance—Equipment		
13-1	Equipment Rental Rates	Ernst & Ernst	22,800
	Area Fourteen Maintenance—Maintenance of Way and Structures		
14-1	Upgrading of Unit Maintenance Cost Index and Development of Interstate Maintenance Requirements	Tallamy Assoc	205,128*
14-2	Techniques for Reducing Roadway Occupancy During Routine Maintenance Activities	—	Contract Pending
	Area Fifteen: Design—General Design		
15-1	Guardrail Design	Cornell Aero Lab	19,723*
15-1(2)	Guardrail Performance and Design	Sw Research Inst	380,000
15-2	Design to Control Erosion in Roadside Drainage Channels	U of Minnesota	95,000
15-3	Rational Structural Analysis and Design of Pipe Culverts	Northwestern U	49,937
15-4	Estimating Runoff Rates from Small Rural Watersheds	Travelers Res Cen	299,902
15-5	Dynamic Characteristics of Heavy Highway Vehicles	Gen Mot Corp	135,000
15-6	Development of Criteria for Safer Luminaire Supports	Texas A & M	150,000
	Area Sixteen: Design—Roadside Development		
16-1	Effects of Deicing Compounds on Vegetation and Water Supplies	V P I	200,000
16-2	Evaluation of Research on Roadside Development	Western States	100,000

PROJECT NO.	TITLE	RESEARCH AGENCY	CONTRACT AMOUNT OR CONTRACT COST
	Area Seventeen: Traffic—Safety		
17-1	Development of Improved Methods for Reduction of Traffic Accidents	Cornell Aero Lab	250,000
	Area Eighteen: Materials and Construction—Concrete Materials		
18-1	Revibration of Retarded Concrete for Continuous Bridge Decks	U of Illinois	106,300
	Area Nineteen: Administration—Finance		
19-1	Budgeting for State Highway Departments	Ernst & Ernst	45,000
19-2(1)	Develop Performance Budgeting System to Serve Highway Maintenance Management	Booz-Allen & Ham.	6,000*
19-2(2)	Develop Performance Budgeting System to Serve Highway Maintenance Management	Ernst & Ernst	6,000*
19-2(3)	Develop Performance Budgeting System to Serve Highway Maintenance Management	Jorgensen & Assoc	6,000*
19-2(4)	Develop Performance Budgeting System to Serve Highway Maintenance Management	Jorgensen & Assoc	220,000
19-3	Economic Effects of Changes in Legal Vehicle Weights and Dimensions on Highways	—	Contract Pending
	Area Twenty: Special Projects		
20-1	Highway Research Information Service	HRB	455,000*
20-2	Research Needs in Highway Transportation	Tallamy–Smith	98,760*
20-3	Optimizing Freeway Corridor Operation Through Traffic Surveillance, Communication, and Control	Texas A & M	419,000 200,540†
20-3A	Optimizing Freeway Corridor Operation Through Traffic Surveillance, Communication, and Control	U of Michigan	480,642 20,000††
20-4	Public Preference for Future Individual Transportation	Chilton Research National Analysts	195,260 83,911
20-5	Synthesis of Information Related to Highway Problems	HRB	200,000
20-6	Right-of-Way and Legal Problems Arising out of Highway Programs	HRB	200,000
20-7	Special AASHO Planning and Design Policies Committee Research	Texas A & M	100,000
20-8	Interactive Graphic Systems for Highway Design	—	Contract Pending

PROJECT NO.	TITLE	RESEARCH AGENCY	CONTRACT AMOUNT OR CONTRACT COST
	Area Twenty-One: Soils and Geology—Testing and Instrumentation		
21-1	Instrumentation for Measurement of Moisture	Res Triangle Inst	34,470
	Area Twenty-Two: Design—Vehicle Barrier Systems		
22-1	Concepts for Improved Traffic Barrier Systems	—	Contract Pending

* Final contract cost.

† NCHRP funds obligated under the $314,340 four-way agreement among the National Academy of Sciences, Michigan Department of State Highways, Wayne County, and the City of Detroit.

†† NCHRP funds obligated under the $70,000 five-way agreement among the National Academy of Sciences, Michigan Department of State Highways, Wayne County, the City of Detroit, and the University of Michigan.

APPENDIX O *Research of the Special Projects Division*

Special Projects	Start-End Dates	Advisory Committee	Principals	Sponsors	Funds
1. Guidelines for Satellite Studies of Pavement Performance	3/63 2/64	J. Havens	P. Irick R. Hudson	NCHRP 1–1	$ 42,800
2. Development of the Highway Research Information Service (HRIS)	3/64 9/67	O. Marzke	P. Irick	BPR NCHRP (20–1)	855,000
3. Improved Street Utilization Through Traffic Engineering	7/66 12/67	F. Hurd	P. Box G. Sessions	AMA, API, ASF	45,000
4. Design and Performance Criteria for Improved Nonrail Transit Vehicles	1/67 3/68	D. Berry	B. Bledsoe	HUD	150,000
5. Evaluation of Driver Education and Training Programs	7/68 3/69	W. Churchman	H. Harman (ETS)	NHSB	120,000
6. Ambulance Design Criteria	6/68 7/69	J. Baerwald	R. Sigafoo	NHSB	64,000
7. National Study of the Composition of Roadside Litter	9/68 8/69		A. Finkner (RTI)	KAB	15,000
8. Development of Pilot Transportation Research Activity Information Services (TRIS I)	12/68 3/70		P. Irick S. Schofer	DOT-OST	59,000
9. Summary and Evaluation of Economic Consequences of Highway Improvements	1/67 12/70 (est.)	M. Snider	R. Winfrey C. Zellner (G. St. Clair)	NCHRP (2–11)	100,000
10. Synthesis of Existing Information Relating to Highway Problems	12/67 Continuing		T. Copas	NCHRP	100,000 per year
11. Right-of-Way and Legal Problems	11/68 Continuing	R. Netherton	J. Vance	NCHRP (20–6)	200,000 per 2-3 years
12. Development of Maritime Research Information Service (MRIS)	6/69 Continuing	G. Simpson	J. Oren (MIC) D. Mellor (MIC) S. Schofer (HRB)	Maritime Administration	39,000 (MIC) 60,000 (HRB)
13. Development of Transportation Noise Research Information Service (TNRIS)	1/70 Continuing	W. Galloway	J. Wright S. Schofer	DOT-OST	116,000
14. Development of Highway Safety Information Service (HSIS)	4/70 Continuing		S. Schofer C. Taylor	NHSB	130,000

APPENDIX P *Conferences, Workshops, and Special Meetings Involving the Highway Research Board*

1971 (*Proposed*)

HRB Western Summer Meeting, "Automation Systems for Highway Organizations"
August 16-18, 1971—University of Texas at Austin
Financial Support: Registration fee
Cosponsors: Texas Highway Department and Center for Highway Research, College of Engineering, University of Texas

Fourth Annual Human Factors Workshop
January 17-18, 1971—Washington, D. C.
Financial Support: Registration fee
Originator: HRB Committee on Road User Characteristics and Committee on Motorist Information Systems
Advisory Committee Chairman: Burton Stephens

International Workshop on Transportation Research Information Transfer
January 17-18, 1971—Washington, D. C.
Financial Support: Registration fee
Originator: HRB Special Projects Division and Special Committee on International Cooperative Activities
Workshop Chairman: Dr. G. W. Cleven

Conference on Organization for Continuing Urban Transportation Planning
April 1971—California (tentative)
Cosponsored and supported by: U. S. Department of Transportation (Federal Highway Administration, Urban Mass Transportation Administration, and Office of the Assistant Secretary for Environment and Urban Systems), and Highway Users Federation for Safety and Mobility
Originator: HRB Committee on Organization and Administration
Ad Hoc Advisory Committee Chairman: Frederick T. Aschman

Workshop on Anti-Skid Program Management
January 17-18, 1971—Washington, D. C.
Financial Support: Registration fee
Originator: Group 3 Council
Steering Committee Cochairmen: J. A. Murchie and E. S. Hunter

1970

Ninth Annual Highway Laws Workshop
July 27-31, 1970—University of North Carolina at Asheville
Financial Support: Registration fee
Cosponsors: Office of Attorney General of North Carolina, North Carolina State Highway Commission, and University of North Carolina at Asheville
Originator: HRB Legal Resources Group

Second Annual Institute on Motor Vehicle and Traffic Law
July 20-23, 1970—University of Colorado, Boulder
Financial Support: Registration fee
Cosponsors: American Association of Motor Vehicle Administrators and University of Colorado Law School
Originator: HRB Committee on Motor Vehicle and Traffic Law

Third National Conference on Rail-Highway Grade Crossing Safety
August 25-27, 1970—Georgia Institute of Technology, Atlanta
Financial Support: Registration fee and National Safety Council
Cosponsors: National Safety Council and Georgia Institute of Technology
Originator: HRB Committee on Highway-Railroad Grade Crossings
Conference Cochairmen: Hoy A. Richards and Douglas M. Fergusson

Workshop on the Impact of the Bay Area Rapid Transit System
February 9-11, 1970—University of California, Berkeley
Financial Support: Registration fee
Cosponsor: Joint Program of Urban Transportation Study, University of California
Originator: HRB Committee on Socioeconomic Aspects of Highways
Conference Planning Committee Chairman: Warren B. Lovejoy
Proceedings were published as Special Report 111

Conference on the Use of Census Data in Transportation Planning
July 9-10, 1970—Washington, D. C.
Originator: HRB Committee on Transportation Forecasting and Committee on Transportation Information Systems
Conference Cochairmen: George V. Wickstrom and Robert E. Barraclough

Symposium on Snow and Ice Control Research
April 8-10, 1970—Dartmouth College, Hanover, New Hampshire
Financial Support: Registration fee
Cosponsors: U. S. Army Cold Regions Research and Engineering Laboratory, with the cooperation of Dartmouth College
Originator: HRB Committee on Snow and Ice Control
Steering Committee Chairman: L. D. Minsk
Proceedings were published as Special Report 115

Second Maintenance Management Workshop
August 3-5, 1970—University of Illinois, Urbana
Financial Support: Registration fee
Cosponsors: Illinois Division of Highways and University of Illinois
Originator: Department of Maintenance
Steering Committee Chairmen: H. O. Scheer and R. L. McCracken

HRB Western Summer Meeting, "Improving Pavement and Bridge Deck Performance"
August 17-19, 1970—Sacramento, California
Financial Support: Registration fee
Cosponsor: California Division of Highways

Workshop on Structural Design of Asphalt Concrete Pavement Systems
December 7-11, 1970—University of Texas
Financial Support: Federal Highway Administration, Bureau of Public Roads
Originator: HRB Committee on Theory of Pavement Design, Committee on Subsurface Drainage, and Committee on Embankments and Earth Slopes
Advisory Committee Chairman: Carl L. Monismith

Conference on Urban Goods Movement
December 7-9, 1970—Airlie House, Warrenton, Virginia
Financial Support and Cosponsors: U. S. and Canadian Departments of Transportation
Originator: HRB Committee on Freight Transportation Economics
Advisory Committee Chairman: J. Douglas Carroll, Jr.

Traffic Conference
February 12-13, 1970—Washington, D. C.
Financial Support: Traffic Education and Training Committee of the National Safety Council
Cosponsor and Host: Highway Research Board

Workshop on Demand-Actuated Transportation Systems
September 10-11, 1970—Purdue University
Financial Support: Registration fee
Cosponsor: Purdue University
Originator: HRB Committee on New Transportation Systems and Technology

Third Annual Human Factors Workshop
January 11, 1970—Washington, D. C.
Financial Support: Registration fee
Originator: HRB Committee on Road User Characteristics
Advisory Committee Chairman: David Schoppert

1969

HRB Western Summer Meeting
August 11-13, 1969—Salt Lake City, Utah
Cosponsor: Utah Department of Highways
Proceedings were published as Special Report 107

International Conference on the Effects of Temperature and Heat on the Engineering Properties of Soils
January 16, 1969—Washington, D. C.
Financial Support: National Science Foundation
Originator: HRB Committee on Physico-Chemical Phenomena in Soils
Proceedings were published as Special Report 103

Second Annual Symposium on Visibility and Driving
July 8-10, 1969—University of California, Berkeley
Cosponsors: Institute of Transportation and Traffic Engineering and Illuminating Engineering Research Institute
Steering Committee Chairman: Matthew C. Sielski
Symposium resulted in a publication

National Conference on Rail-Highway Grade Crossing Safety
February 11-13, 1969—University of Illinois, Urbana
Financial Support: U. S. Department of Transportation
Cosponsors: Highway Traffic Safety Center and Division of University Extension, University of Illinois
Originator: HRB Committee on Highway Safety
Program Planning Committee Chairman: Hoy A. Richards

Conference on Transportation and Community Values
March 2-5, 1969—Airlie House, Warrenton, Virginia
Financial Support: Bureau of Public Roads, U. S. Department of Housing and Urban Development, with cooperation from Urban Mass Transportation Administration
Advisory Committee Chairman: Frederick T. Aschman
Originator: HRB Department of Urban Transportation Planning
Proceedings were published as Special Report 105

Conference on Relocation—Social and Economic Aspects
October 1-3, 1969—University of Maryland, College Park
Originator: HRB Committee on Socio-Economic Aspects of Highways
Conference Planning Committee Chairman: Floyd I. Thiel
Proceedings were published as Special Report 110

Conference on Transportation of Hazardous Materials
May 6-9, 1969—Airlie House, Warrenton, Virginia
Financial Support: U. S. Department of Transportation, Office of Hazardous Materials
Cosponsors: NAS Committee on Hazardous Materials
Originator: HRB Department of Economics, Finance and Administration
Steering Committee Chairman: D. L. Katz

Second Annual Human Factors Workshop
January 12, 1969—Washington, D. C.
Financial Support: Registration fee
Originator: HRB Committee on Road User Characteristics
Advisory Committee Chairman: Charles Baker

Eighth Annual Highway Laws Workshop
July 28-31, 1969—University of Washington, Seattle
Financial Support: Registration fee
Cosponsors: University of Washington School of Law and Office of Short Courses and Conferences and the Office of the Attorney General, State of Washington
Originator: HRB Department of Legal Studies
Planning Committee Chairman: John P. Holloway

Institute on Motor Vehicle and Traffic Law
August 19-21, 1969—University of Colorado, Boulder
Financial Support: Registration fee
Cosponsor: American Association of Motor Vehicle Administrators and University of Colorado Law School
Originator: HRB Committee on Motor Vehicle and Traffic Law
Program Chairman: Louis R. Morony

1968

HRB Western Summer Meeting, "Control of Pavement Slipperiness and Asphalt Pavement Cracking"

August 12-13, 1968—Denver, Colorado
Financial Support: Registration fee
Cosponsor: Colorado Department of Highways
Proceedings were published as Special Report 101

Pilot Workshop on Human Factors in the Design and Operation of the Highway Transportation System

January 14, 1968—Washington, D. C.
Financial Support: Automotive Safety Foundation and Matrix Corporation
Originator: HRB Road User Characteristics Committee
Pilot Workshop Committee Chairman: J. E. Uhlaner

Seventh Annual Highway Laws Workshop

July 29-August 1, 1968—University of Nevada, Reno
Financial Support: Registration fee
Cosponsors: Nevada State Highway Department and University of Nevada
Originator: HRB Department of Legal Studies

Maintenance Management Workshop

July 22-24, 1968—Ohio State University, Columbus
Financial Support: Automotive Safety Foundation (assistance with publication)
Cosponsor: Department of Civil Engineering, Ohio State University
Originator: HRB Department of Maintenance and Department of Economics, Finance and Administration
Advisory Committee Chairman: Roy E. Jorgensen
Proceedings were published as Special Report 100

Symposium on Visibility in the Driving Task

May 13-15, 1968—Texas A&M University, College Station
Financial Support: Registration fee
Cosponsors: Illuminating Engineering Research Institute and Texas A&M University
Originator: HRB Committee on Night Visibility
Advisory Committee Chairman: Matthew C. Sielski
Symposium resulted in a publication

Conference on Joint Development and Multiple Use of Transportation Rights-of-Way

November 14-15, 1968—Washington, D. C.
Financial Support: American Association of State Highway Officials, U. S. Department of Housing and Urban Development, U. S. Department of Transportation, and financial assistance from Automotive Safety Foundation for publication of the proceedings
Originator: ITTE and Civil Engineering Department of Polytechnic Institute of Brooklyn
Conference Steering Committee Chairman: D. Grant Mickle
Proceedings were published as Special Report 104

Symposium on Wearing Surfaces for Steel Bridge Decks
September 7, 1968—Columbia University, New York
Sponsor: IABSE
Cosponsors: Highway Research Board, American Association of State Highway Officials, and American Society of Civil Engineers
Organizing Committee Chairman: Dr. Ivan Viest
Symposium resulted in a publication

Engineering Foundation Research Conference on Access to Airports
July 29-August 2, 1968—University School, Milwaukee, Wisconsin
Cooperating Organizations: Highway Research Board, U. S. Department of Transportation, U. S. Department of Housing and Urban Development, American Transit Association, and the Highway Division of the American Society of Civil Engineers

1967

Conference on Improved Street Utilization Through Traffic Engineering
May 22-24, 1967—Washington, D. C.
Cosponsors: Automotive Safety Foundation, Automobile Manufacturing Association, and American Petroleum Institute
Originator: HRB Department of Traffic and Operations
Advisory Committee Chairman: Fred W. Hurd
Proceedings were published as Special Report 93

Conference on Urban Development Models
June 26-30, 1967—Dartmouth College, Hanover, New Hampshire
Financial Support: Bureau of Public Roads, Automotive Safety Foundation, and U. S. Department of Housing and Urban Development
Originator: HRB Committee on Land Use Evaluation
Advisory Committee Chairman: Britton Harris
Proceedings were published as Special Report 97

Sixth Annual Workshop on Highway Laws
July 24-27, 1967—University of Wisconsin, Madison
Financial Support: Registration fee
Cosponsors: University of Wisconsin Law School and State Highway Commission of Wisconsin
Originator: HRB Department of Legal Studies
Papers published in Highway Research Record 260

1966

Fifth Annual Workshop on Highway Law
July 11-15, 1966—University of Colorado, Boulder
Cosponsors: University of Colorado Law School and State Highway Department of Colorado
Originator: HRB Department of Legal Studies
Papers published in Highway Research Record 166 and 260

1965

Fourth Annual Highway Laws Workshop
July 12-16, 1965—Washington University, St. Louis, Missouri
Cosponsors: Washington University School of Law and Missouri State Highway Commission
Originator: HRB Department of Legal Studies

A Colloquy on Motor Vehicle and Traffic Law
February 2-3, 1965—Washington, D. C.
Financial Support: Automotive Safety Foundation
Originator: HRB Committee on Motor Vehicle and Traffic Law
Proceedings were published as Special Report 86

Correlation and Conference of Portable Nuclear Density and Moisture Systems
July 12, 1965—Charlottesville, Virginia
Financial Support: U. S. Bureau of Public Roads through HPR funds available to Virginia Department of Highways
Cosponsors: Virginia Council of Highway Investigation and Research and Highway Research Board
Originator: HRB Special Committee on Nuclear Principles and Applications and American Society for Testing and Materials Committee E-10, Subcommittee IV on Measurement Using External Radiation Sources
Proceedings were published in Highway Research Record 177

1964

Third Annual Workshop on Highway Laws
April 13-17, 1964—Louisiana State University, Baton Rouge
Financial Support: Registration fee
Cosponsors: Louisiana Department of Highways and Louisiana State University Law School
Originator: HRB Department of Legal Studies
Papers published in Highway Research Record 260

Course on Radioisotope Applications to Highway Engineering
March 9-27, 1964—Oak Ridge, Tennessee
Financial Support: Registration fee
Cosponsor: U. S. Atomic Energy Commission, with the cooperation of the Oak Ridge Institute of Nuclear Studies
Originator: HRB Special Committee on Nuclear Principles and Applications
Conference Coordinator: W. G. Gunderman

1963

International Permafrost Conference
November 11-15, 1963—Purdue University, Lafayette, Indiana
Financial Support: U. S. Bureau of Public Roads, Caterpillar Tractor Company, National Science Foundation, U. S. Office of Civil Defense, Office of Naval Research, Purdue University School of Civil Engineering, U. S. Air Force Cambridge

Research Laboratories–Terrestrial Science Laboratory, U. S. Army Corps of Engineers, U. S. Army Materiel Command, U. S. Navy Facilities Engineering Command, and U. S. Public Health Service

Cosponsors: American Geophysical Union; American Society of Civil Engineers, Soil Mechanics and Foundations Division; American Society for Testing and Materials, Division of Materials Sciences; National Academy of Sciences–National Research Council, Highway Research Board; and National Research Council of Canada, Associate Committee on Snow and Soil Mechanics

Originator: Building Research Advisory Board

General Conference Chairman: Kenneth B. Woods

Second Annual Workshop on Highway Laws

June 24-28, 1963—University of Wisconsin, Madison

Financial Support: Registration fee

Cosponsors: U. S. Bureau of Public Roads and University of Wisconsin

Originator: HRB Committee on Highway Laws

Papers published in Highway Research Record 258 and 260

1962

Workshop on Highway Laws

July 15-20, 1962—University of Wisconsin, Madison

Cosponsors: U. S. Bureau of Public Roads and University of Wisconsin Extension Law Department

Originator: HRB Special Committee on Highway Laws

Papers published in Special Report 76

Conference on Urban Transportation Research, "A Key to Change"

January, 1962—Washington, D. C.

Originator: HRB Committee on Urban Transportation Research

Proceedings were published as Special Report 69

AASHO Road Test Conference

May 16-18, 1962—St. Louis, Missouri

Financial Support: American Association of State Highway Officials

National Advisory Committee Chairman: K. B. Woods

Proceedings were published as Special Report 73

Conference on Planning in Highway Administration

March 26-27, 1962—Washington, D. C.

Financial Support: Bureau of Public Roads

Originator: HRB Committee on Highway Organization and Administration

Proceedings were published as Special Report 72

1960

Interim Study Group for Appraising Instrumentation and Procedure for Determination of Deflection Characteristics of Highway Surfaces Under Moving Loads

September 20, 1960—Washington, D. C.

Originator: HRB Department of Materials and Construction

Study Group Chairman: Frank P. Nichols, Jr.

Workshop on Formulating Highway Construction Programs
September 19-20, 1960—Washington, D. C.
Financial Support: Automotive Safety Foundation and U. S. Bureau of Public Roads
Originator: HRB Department of Economics, Finance and Administration
Proceedings were published as Special Report 62

1959

Conference on Economic Analysis in Highway Programming
September 17-18, 1959—Washington, D. C.
Financial Support: Automotive Safety Foundation and U.S. Bureau of Public Roads
Originator: HRB Committee on Economic Analysis
Proceedings were published as Special Report 56

Soil and Foundation Engineering in the USSR
September 14-October 5, 1959, and May 31-June 21, 1959—Washington, D. C.
Financial Support: National Science Foundation
Cooperating Organizations: American Society of Civil Engineers, Soil Mechanics Division; U. S. National Committee of International Society of Soil Mechanics and Foundation Engineering; and Highway Research Board

1958

An International Symposium on Water and Its Conduction in Soils
January, 1958—Washington, D. C.
Originator: HRB Committee on Physico-Chemical Phenomena in Soils
Chairman: Dr. Hans F. Winterkorn
Proceedings were published as Special Report 40

First International Skid Prevention Conference
September 8-12, 1958—University of Virginia
Sponsor: University of Virginia
Cooperating Organizations: Highway Research Board and others

National Conference on Highway and Urban Development
October 5-9, 1958—Sagamore Conference Center, Syracuse University
Financial Support: Automotive Safety Foundation
Cosponsors: U.S. Bureau of Public Roads, Joint Committee on Highways of American Municipal Association and the American Association of State Highway Officials, and Syracuse University
Originator: HRB Committee on Urban Research

1957

Conference on Economic Impact of Highway Improvement

March 18-19, 1957—Washington, D. C.

Financial Support: Automotive Safety Foundation and U. S. Bureau of Public Roads

Originator: American Association of State Highway Officials by Resolution to Highway Research Board

Conference Chairman: David R. Levin, Chairman of HRB Committee on Land Acquisition and Control of Highway Access and Adjacent Areas

Proceedings were published as Special Report 28

Index

A

D

E

F

G

H

I

J

K

L

M

N

O

P

R

S

T

U

V

W